Sustainability Studies: Environmental and Energy Management

Edited by

G. Venkatesan

Department of Civil Engineering, University College of Engineering (BIT Campus), Anna University, India

S. Lakshmana Prabu

Department of Pharmaceutical Technology, University College of Engineering (BIT Campus), Anna University, India

&

M. Rengasamy

Department of Petrochemical Technology, University College of Engineering (BIT Campus), Anna University, India

Sustainability Studies: Environmental and Energy Management

Editors: G. Venkatesan, S. Lakshmana Prabu and M. Rengasamy

ISBN (Online): 978-981-5039-92-4

ISBN (Print): 978-981-5039-93-1

ISBN (Paperback): 978-981-5039-94-8

need for a court order if at any point you breach any terms of this License Agreement. In no event will any delay or failure by Bentham Science Publishers in enforcing your compliance with this License Agreement constitute a waiver of any of its rights.

3. You acknowledge that you have read this License Agreement, and agree to be bound by its terms and conditions. To the extent that any other terms and conditions presented on any website of Bentham Science Publishers conflict with, or are inconsistent with, the terms and conditions set out in this License Agreement, you acknowledge that the terms and conditions set out in this License Agreement shall prevail.

Bentham Science Publishers Pte. Ltd.
80 Robinson Road #02-00
Singapore 068898
Singapore
Email: subscriptions@benthamscience.net

BENTHAM SCIENCE

CONTENTS

PREFACE

The purpose of this book, "Sustainability Studies: Environmental and Energy Management" is to provide a continuous state-of-the-art critical view of the current knowledge of the environment and energy management. Recent Global warming, heat around the earth, pollution, unpredicted atmosphere changes, and contaminants such as physical and biological components severely affect the normal environment. There are no boundaries observed for these environmental stressors and pollution. Recently these two factors have been considered to have a significant role environmental as well as energy management.

This book covered various topics related to the fundamental nature of contaminants, their measurements, characterization and different techniques for their removal.

With this idea, we have chosen authors from different countries with varied backgrounds who could add the issues and solutions which provide the readers an understanding of global issues in a comprehensive better manner. In the beginning, a country or a person who has an interest in environmental and energy management can build a platform. It is imperative that we can build a strong management system globally.

The diversity of authors helps us disclose the advances in environmental and energy management worldwide. Their inputs will let the readers know about the problems and issues faced by academic individuals, industries, and research institutions and the directions for future prospects.

This book presents a complete lookout to provide a source of information on all facets of environmental and energy management for undergraduate and post-graduate students, researchers both industrial and academicians in the field of environmental and energy management.

As the editors of this book, we are sure that the book chapters surely guide the students and researchers with unexplored avenues and future perspectives in environmental and energy management to identify promising solutions that might represent future solutions in these critical areas.

Jeeva Chithambaram authored a chapter titled "Alternative building materials-Road to Sustainability" where they discussed the building materials which are made from industrial waste as conventional building materials for sustaining the environment from degradation.

The chapter titled "preparation of environmental friendly thin films using SILAR method" by mani describes how thin films are prepared by the SILAR method, adopting some simple adsorption and reaction of the ions from the solution as a chemical method.

Shenbagavadivu has written a chapter on "Smart waste management to enrich cleanliness and reduce pollution in environment". They conversed about the importance of solid waste management to safeguard public health by adopting a smart solid waste management system for garbage collection to enhance environmental sustainability which can render support to the economic growth of the Nation.

The chapter titled "A new philosophy of production" is written by Markic. It discusses the growth and development of society on our planet by the consumption of natural resources which lead to increase the production of waste and other substances which are harmful for both human beings as well as the ecosystem. Considering the several factors such as circular economy, industrial ecology, ecological economy, blue economy, biomimicry, cradle to cradle, cleaner production and regenerative design, they discussed the approach to production to ensure that man functions in accordance with natural laws, and that we need to leave nature and the environment in a much better condition than we inherited.

Lakshmana Prabu and Thirumurugan authored a chapter titled "Remediation approaches for the degradation of textile dye effluents as sustaining environment". They have deliberated on the impact of textile dye effluents and different remediation approaches in order to maintain the sustainability of this environment in near future.

The chapter titled "A comparative adsorption study of acid violet 7 and brilliant green dyes in aqueous media using rice husk ash (RHA) and coal fly ash (CFA) mixture" is written by Irvan. They demonstrated the adsorption of dyes such as acid violet 7 (AV7) and brilliant green (BG) in the mixture of rice husk ash (RHA) and coal fly ash (CFA) as adsorbents and highlighted that RHA-CFA can act as a very good adsorbent to remove AV7 and BG from aqueous medium.

Ashok Kumar discusses the "Pollution Prevention Assessments: Approaches and Case Histories" in Ohio, USA. This chapter focuses on pollution prevention by applying the concept of energy efficiency, energy savings, greenhouse gas emission (GHG) reductions, waste reduction, and stormwater management.

The chapter "Threats to sustainability of land resources due to aridity and climate change in the northeast agro-climatic zone of Tamil Nadu, South India" by Dhanya is based on the Aridity Index.

It was a great experience to write and edit this book. We acknowledge the full support of all the authors and specially thank particularly the support extended by Bentham Science and their team had been extremely supportive in the completion of this book.

Dr. G. Venkatesan
Department of Civil Engineering
University College of Engineering (BIT Campus)
Anna University
India

Dr. S. Lakshmana Prabu
Department of Pharmaceutical Technology
University College of Engineering (BIT Campus)
Anna University
India

&

Dr. M. Rengasamy
Department of Petrochemical Technology
University College of Engineering (BIT Campus)
Anna University
India

List of Contributors

A. Ramachandran   Centre for Climate Change and Adaptation Research, Anna University, Chennai, India

Abhijeet Singh   Department of Mechanical and Automation Engineering, Amity University Jharkhand, Ranchi, Jharkhand, India

Anbarasi Karunanithi   University College of Engineering, BIT Campus, Anna University, Trichy - 620 024, India

Andi Mulkan   Mechanical Engineering Study Program, Faculty of Engineering, University of Iskandar Muda, Jalan Kampus Unida - Surien, Banda Aceh 23234, Indonesia

Ashok Kumar   College of Engineeringf, The University of Toledo, Toledo, Ohio, USA 43606

B.P. Naveen   Department of Civil Engineering, Amity University Haryana, India

C. Ravichandran   Department of Environmental Sciences, Bishop Heber College, Tiruchirappalli, Tamilnadu, India

Dhanaraja Dhanapal   Paavai Engineering College, Pachal, Namakkal - 637018, India

Dragana Nešković Markić   Pan-European University "Apeiron", Banja Luka, Republic of Srpska, Bosnia and Herzegovina

G. Jenilasree   Reaserach Scholar, University College of Engineering, BIT Campus, Anna University, Tiruchirappalli, India

G. Venkatesan   Department of Civil Engineering, University College of Engineering (BIT Campus), Anna University, Tiruchirappalli 620 024, India

G.S. Rampradheep   Department of Civil Engineering, Kongu Engineering College, Perundurai, Tamilnadu, India

Haider M. Zwain   College of Water Resources Engineering, Al-Qasim Green University, Al-Qasim Province, Babylon, Iraq

Irvan Dahlan   School of Chemical Engineering, Universiti Sains Malaysia, Engineering Campus 14300 Nibong Tebal, Pulau Pinang, Malaysia

K. Palanivelu   Centre for Climate Change and Adaptation Research, Anna University, Chennai, India

K.S. Divya   VTU Extension Center, Karnataka State Remote Sensing Applications Centre, Bangalore, India

Lakshika Nishadhi Kuruppuarachchi   College of Engineeringf, The University of Toledo, Toledo, Ohio, USA 43606

Ljiljana Stojanović Bjelić   Pan-European University "Apeiron", Banja Luka, Republic of Srpska, Bosnia and Herzegovina

M. Bhuvaneswari   Department of ECE, University College of Engineering, BIT Campus, Anna University, Tiruchirappalli, India

M. Rengasamy   Department of Petrochemical Technology, University College of Engineering (BIT Campus), Anna University, Tiruchirappalli 620 024, India

N. Shenbagavadivu   Department of Computer Applications University College of Engineering, BIT Campus, AnnaUniversity, Tiruchirappalli, India

P. Dhanya   Centre for Climate Change and Adaptation Research, Anna University, Chennai, India

Pardeep Bishnoi   Associate IP, Clarivate Analytics, Noida, India

Predrag Ilić   Institute for Protection and Ecology of the Republic of Srpska, Banja Luka, Bosnia and Herzegovina

R. Thirumurugan   College of Agriculture and Life Sciences, North Carolina State University, Kannapolis, NC, USA

S. Jeeva Chithambaram   Department of Civil Engineering, Sarala Birla University, Ranchi, Jharkhand, India

S. Lakshmana Prabu   Department of Pharmaceutical Technology, University College of Engineering (BIT Campus), Anna University, Tiruchirappalli, India

Saisantosh Vamshi Harsha Madiraju   College of Engineeringf, The University of Toledo, Toledo, Ohio, USA 43606

Sariyah Mahdzir   School of Civil Engineering, Universiti Sains Malaysia, Engineering Campus 14300 Nibong Tebal, Pulau Pinang, Malaysia

Zarook Shareefdeen   Department of Chemical Engineering, American University of Sharjah, P.O. Box 26666, Sharjah, UAE

CHAPTER 1

A New Philosophy of Production

Dragana Nešković Markić[1,*], Predrag Ilić[2] and Ljiljana Stojanović Bjelić[1]

[1] *Pan-European University "Apeiron", Banja Luka, Republic of Srpska, Bosnia and Herzegovina*

[2] *Institute for Protection and Ecology of the Republic of Srpska, Banja Luka, Bosnia and Herzegovina*

Abstract: The growth and development of society on our planet has caused a great consumption of natural resources and, on the other hand, the production of waste and other substances harmful both to human health and to the ecosystem itself. With this way of life, man has moved away from nature. Consequently, a system that functions contrary to natural laws has been established. With the new way of production, it is necessary to return to natural processes and sustainable technologies, clean technologies, and the use of renewable energy sources. The projection of sustainability in the future must be based on resource use restriction, material reuse and other principles of economic and environmental sustainability. This chapter will discuss the new approach to production and the product itself through the consideration of several different possibilities such as circular economy, industrial ecology, ecological economy, blue economy, biomimicry, cradle to cradle, cleaner production and regenerative design. The above-mentioned possibilities in production, design and the product itself aim to ensure that man functions in accordance with natural laws, and that we need to leave nature and the environment in a much better condition than we inherited.

Keywords: Biomimicry, Blue Economy, Circular Economy, Cleaner Production, Cradle to Cradle, Industrial Ecology, Regenerative Design.

INTRODUCTION

A new philosophy of production in the 21st century, or sustainable production, is to create goods and services using processes and technologies that will not create pollution, save energy as well as reduce the pressure on resource depletion. In addition, this production must be economically sustainable, safe, and secure for employees, the local community, and consumers. Sustainable production should reduce the consumption of raw materials and energy per unit of product, as well as improve the quality of the environment and social well-being [1].

* **Corresponding author Dragana Nešković Markić:** Pan-European University "Apeiron", Banja Luka, Republic of Srpska, Bosnia and Herzegovina; Tel: +38751247923; E-mail: dragana.d.neskovicmarkic@apeiron-edu.eu

G. Venkatesan, S. Lakshmana Prabu and M. Rengasamy (Eds.)

Since the 1970s, the management and control of pollutant emissions have been based on the exhaust pipes or chimneys, the so-called "end-of-pipe" technology. In any case, this approach to environmental pollution control has yielded results in many aspects, but the regulations have not been aimed at preventing pollution or impacts in the future. New technologies deal with the cause of the problem, as opposed to "end-of-pipe" technologies that deal with symptoms. It often happened that pollution was transferred from one environmental medium to another (water, air, land) or transferred to other geographical areas and other countries. This has resulted in an increasing search to solve problems by changing the production process, rather than treating pollution at the end of the production process.

If we look at the specific strategies of companies, we can determine the focus on disposal and recycling technologies, which are mainly "end-of-pipe" technologies. These technologies are supplemented to the original production process without introducing major changes in the technical system, where the existing technology is supplemented with new components in order to avoid or reduce the negative impact on the environment. So, we have a supplementation of the existing technological process with a filtration and purification plant, disposal method, and recycling technology [2]. "End-of-pipe" technologies essentially continue production without changing the existing technical system, *i.e.*, stabilizing the existing technological system with correcting possible negative impacts on the environment [3].

CIRCULAR ECONOMY

Beginning with the industrial revolution, the global economy is characterized by a significant model, the so-called linear model of production and consumption. The linear model of economics operates on the principle of "take, make, use and discard", according to which all products that man no longer needs end up as waste [4]. It is a well-established practice that products are created at low prices, used and discarded. This approach in a linear economy results in unsustainable extraction of natural resources and the accumulation of pollution [5]. If resource consumption continues at this pace, by 2050, it would take the equivalent of over two planets to support this development and it would not be possible to achieve the better standard of living it aspires to. This approach, from the aspect of economics, has initiated deliberation about the unsustainability of such a model, that is, the unsustainability of modern civilization. On the other hand, the resources on Earth are not infinite and are becoming increasingly endangered. The growth and development of technology have led to a reduction in production and sales prices, to the growth of the living standard of the population, but also to an uncontrolled imbalance between the economic and ecological systems [6]. At the end of the 1970s, we moved toward the direction of extending the life of products as well as reducing the amount of waste through the introduction of a new model of the so-called circular economy.

Geissdoerfer *et al.* [7] define "the Circular economy as a regenerative system in which resource input and waste, emission, and energy leakage are minimized by slowing, closing, and narrowing material and energy loops". The circular economy, with its 3R principles of reducing, reusing and recycling material clearly illustrates the strong linkages between the environment and economics [8]. The circular economy strives to operate on a "product-waste-product" principle, *i.e.* to ensure sustainable resource management, extend product life, reduce waste and use renewable energy sources. With this new approach, waste is almost non-existent, *i.e.* it has been reduced to a minimum.

As shown in Fig. (**1**), at each stage in the circular economy, the aim is to reduce costs and dependence on natural resources, as well as to reduce the amount of waste. This model replaces the "end of life" concept with restoration, the use of renewable energy sources, and the elimination of the use of toxic chemicals. The value of products and materials is maintained for as long as possible and waste is minimized. The product is produced, used, and after reaching its end *i.e.* its service life, the resources it contains are reused, that is the process returns to the beginning. For example, factory waste becomes valuable in another process, products can be repaired, reused or upgraded instead of being thrown away.

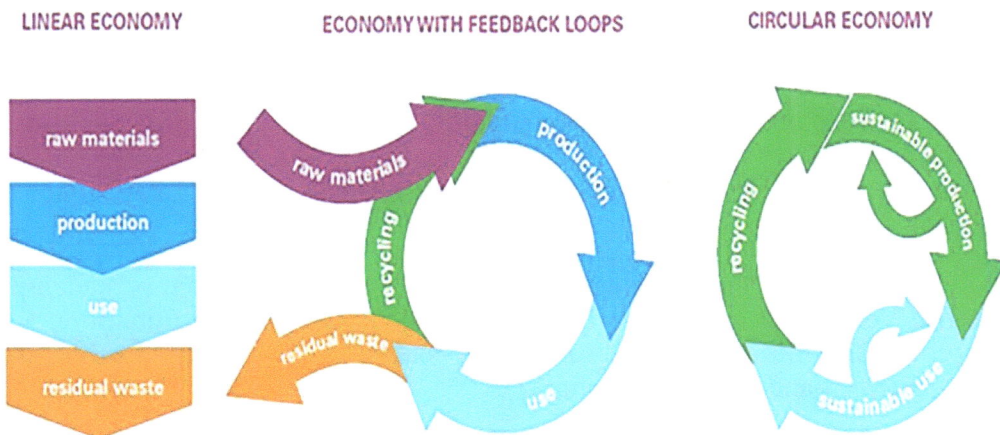

Fig. (1). Linear and circular economy [9].

The circular economy aims to increase the quality of life of citizens, with more efficient use of resources, increased competitiveness, creation of new jobs through the development of new technologies, innovations, designs, and modular products produced in a way that can be easily supplemented and processed, and a new way of organizing in companies. Companies that accept the circular model of the economy imply business in terms of benefits not only for society and the environment but also for consumers and investors themselves.

The linear "take-and-dispose" model relies on large amounts of readily available resources and energy and as such is increasingly unsuitable for the real environment where it operates. In the long run, the costs of solving various environmental problems are increasing, and large amounts of resources are consumed leading to the lack of individual resources [10]. The circular economy is moving toward "zero waste" and changes in product and packaging design and encouraging the idea that waste can become a resource again. The transition from a linear to a circular economy is a strive for all economic actors to spend as few natural resources as possible, rather than going in the direction of regeneration through recycling, the introduction of innovative technologies, *etc*.

The transition from a linear to a circular economy requires changes in the following segments [11]:

- Organization of the company,
- Education,
- Innovation in technologies,
- Creating an appropriate institutional framework and infrastructure,
- Designing new products,
- Design, implementation and development of new business and market models,
- Development of waste management system,
- Changing consumer habits and encouraging the development of new forms of behavior,
- Development of new products that support the concept of the circular economy,
- Defining and proclaiming new policies.

We view the circular economy as a complex holistic approach that is made up of several development directions: Industrial Ecology, Ecological Economy, Blue Economy, Biomimicry, Cradle-to-Cradle, Cleaner Production, Regenerative Design, Performance Economy (Fig. **2**).

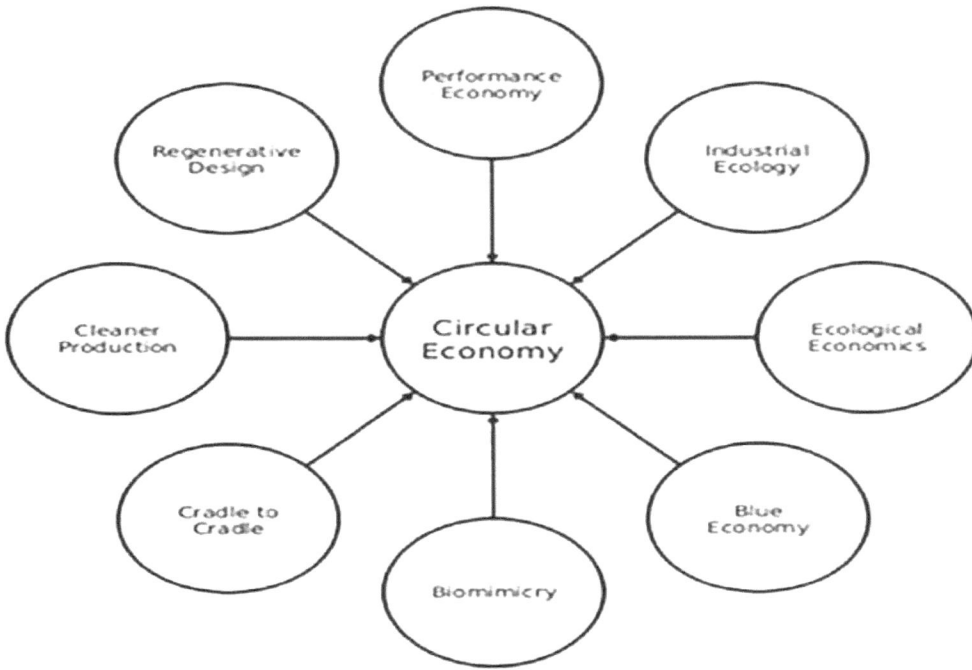

Fig. (2). Key terms closely related to the circular economy [12].

A good example of the application of the circular economy is the Dutch textile company Dye Coo. As is well known, the textile industry uses large amounts of water and chemicals, especially in countries where a large number of these factories are located, such as China, India, Bangladesh, Vietnam and Thailand. Textile company Dye Coo has developed a new innovative technology for dyeing textile materials with carbon dioxide using high pressure. With this dyeing technology, the dye penetrates very well to the depth of the fabric, affecting the non-use of water and other chemicals. Carbon dioxide evaporates and is recycled again. This technology is highly efficient and completely circular. Carbon dioxide is a by-product of other industrial processes and is used again in other dyeing series [13].

Industrial Ecology

Industrial ecology arose from the desire to better understand the impacts of industrial systems on the environment. The first step in industrial ecology is to identify the impacts of industrial systems on the environment and the second step is to implement measures to reduce and minimize these impacts. Industrial ecology can be defined as a scientific field that studies physical, chemical and biological interactions within and between industrial and ecological systems.

The industrial symbiosis is an important aspect of industrial ecology such that wastes, by-products and energy exchanges between different industries for sustainability are practiced. Industrial ecology was established and implemented by developed countries so as to achieve satisfactory levels of sustainability [14].

The industrial ecology is a new approach to industrial product and process design through the application of sustainable production strategies, *i.e.* the industrial system is not observed separately from the surrounding systems. Industrial ecology optimizes the overall material cycle from raw materials to finished materials, products, waste and ultimately disposal.

Research in the field of industrial ecology includes the following activities [15]:

- A systematic approach to the interaction between industrial and ecological systems,
- Study of energy and matter flows, their transformation into products and by-products, waste materials through industrial and natural systems (industrial metabolism),
- Multidisciplinary approach,
- Orientation towards the future,
- Changes from linear (open) processes to closed or cyclic processes where waste from one sector is used as a raw material or input to another process,
- Reducing the impact of industrial systems on natural ecosystems,
- Impact on the harmonious integration of industrial activities into ecological systems,
- Ideas for creating an industrial ecosystem based on more efficient and sustainable natural ecosystems,
- Identification and comparison of industrial and natural systems, which indicate the area of potential studies and activities.

Application of Industrial Ecology

Practical application of industrial ecology–KalundborgEco-Industrial Park in Denmark. The Eco-Industrial Park is an "industrial symbiosis" between numerous companies in and around the city of Kalundborg (Denmark), which has been developing over a period of 25 years. This project is not the result of systematic planning, but was gradually realized and upgraded through the cooperation of surrounding companies. This Eco-Industrial Park is based on the exchange of waste and energy between a thermal power plant, an oil refinery, a pharmaceutical factory and a gypsum factory. Excess heat from the thermal power plant is distributed to households and fish farms. Biowaste from fish farms is used as a fertilizer in agriculture, while excess steam from the power plant is used in a pharmaceutical factory. The gypsum factory uses SO_2 from the waste gases of the thermal power plant. Part of the solid waste is used in the construction of roads [15]. Fig. (3) presents the functioning of the Eco-Industrial Park in Denmark.

Analytical tools used in industrial ecology are [16]:

- Material flow analysis/Substance flow analysis
- Life-cycle assessment, and
- Environmental design

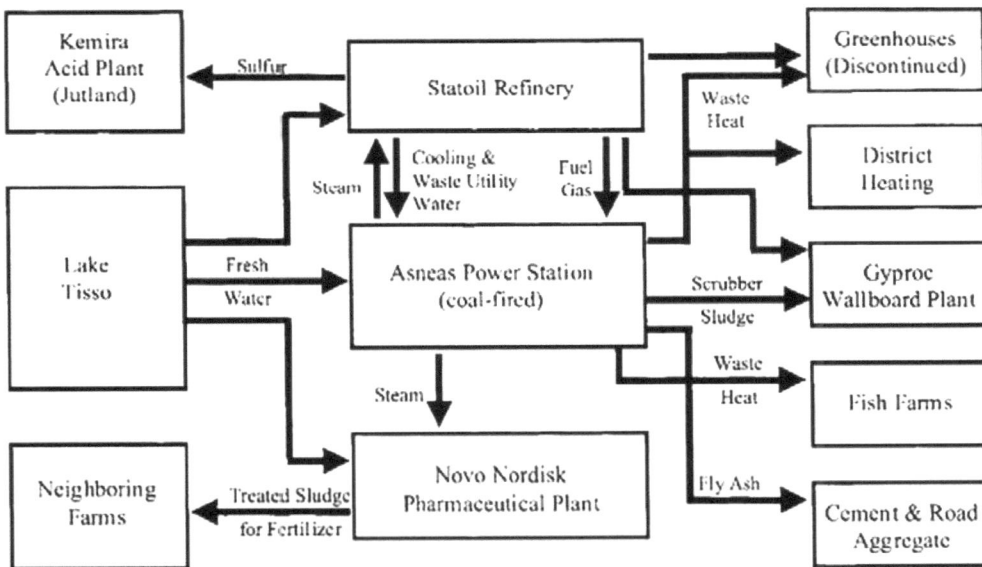

Fig. (3). Scheme of the Eco-Industrial Park Kalundborg, Denmark [15].

Analysis of Material Flows

Material Flow Analysis (MFA) is a method used to describe, investigate and evaluate the metabolism of anthropogenic and geogenic systems. It is a typical analytical tool based on material balance [17]. The basic characteristic of MFA is that the materials behave in accordance with the law of mass maintenance, *i.e.* everything that enters the system must also leave the system. In other words, the matter cannot disappear, it can only transform and leave the system in the form of emissions or other by-products.

MFA is a systematic assessment of material flows and stocks within a system defined in space and time [18]. It connects the sources, paths and transitional or final dispositions of the material. MFA can be implemented at two levels, the level of substances and the level of goods. If we are working at the level of substances, then we are talking about *Substance Flow Analysis* (SFA), and MFA regulates the level of goods, *i.e.* materials.

The first basic principles of MFA preservation of matter, *i.e.* that the input is equal to the output, were postulated by Greek philosophers 2000 years ago. One of the first reports on the analysis of material flows dates back to the 17th century. The Venetian physician SantorioSantorio (1561-1636) researched human metabolism, that is, he was the first to develop a new system balance method 400 years ago. The French chemist Antoine Lavoisier (1743-1794), 200 years after Santorio, proved that the total mass of matter does not change by chemical processes [19, 20].

Brunner and Rechberger [20] presented the terms defined by the MFA methodology: substance, product, material, process, flows and fluxes, stocks, transfer coefficients, system, and system boundaries.

The substance is defined as the material from which goods or products are composed. They can be atoms (C, Pb) or compounds (CO_2, H_2O). Products are defined as substances or mixtures of substances (wood, waste, automobile).

Process means the transport, transformation, storage, or change in the value of a substance or product. The process can be an activity (incineration), plant (incinerator, landfill, composting plant), service (waste collection), or environmental medium (atmosphere, hydrosphere, soil) [21]. Processes are related to flows (mass per unit time) or fluxes (mass per unit time through a unit area). Flows or fluxes of materials entering the process are called input streams, while those leaving the process are called output streams.

Due to the law of conservation of matter, MFA results can be controlled by a simple material balance, comparing all inputs, stocks, and process results [22]. The balance between products and/or substances throughout the process taking into account inputs and outputs is shown in the following formula:

$$\sum_{n=1}^{i} \text{inputs } i = \sum_{n=1}^{j} outputs\ j + \Delta stocks \qquad\qquad \textbf{(1)}$$

Where stocks are defined as the accumulation or degradation of materials in the process.

System boundaries play an important role in MFA design, because the processes within the system must be balanced, and must be defined in space and time.

MFA can be applied at different levels. It can be applied at the international, national, regional, community level, enterprise level, as well as at anthropogenic systems.

A large number of studies for SFA have been analyzed for a large number of substances, for different geographical areas and under time constraints. Analysis of phosphorus flow in Switzerland was investigated by Brunner *et al.* [23], and in China by Li *et al.* [24] and Yuan *et al.* [25]. The metabolism of the city of Paris and its region was investigated by Barles [26].

Life Cycle Assessment

In the late 1960s, there was a need for more responsible management of industrial processes, as there was a growing awareness that resources were limited as well as the growing pressure of environmental pollution. Therefore, the development of a tool that would deal with the assessment of the environmental impact of products on the environment - Life Cycle Assessment (LCA), was approached. LCA is a tool for making decisions about the manufacture or quality of a product with the identification of its impact on the environment, having in mind the entire life cycle of the product, *i.e.* the process of analyzing materials, energy, emissions and waste "produced" by the product, throughout the life cycle from the beginning, *i.e.* starting from the resources and exploitation of materials to the final disposal [27]. Life cycle analysis has proven to be a useful tool in the waste management sector.

The Society of Environmental Toxicology and Chemistry (SETAC) was the first organization to initiate the harmonization of the LCA methodology, which was later grouped into ISO standards. The ISO standard [28] and the ILCD manual (The International Reference Life Cycle Data System) [29] regulate LCA implementation in more detail. ISO 14040 [28] defines four steps in an LCA study (Fig. **4**).

- Defining the goal and subject,
- Life cycle inventory analysis (data collection phase),
- Assessment of the impact of the life cycle on the environment,
- Interpretation of life cycle results.

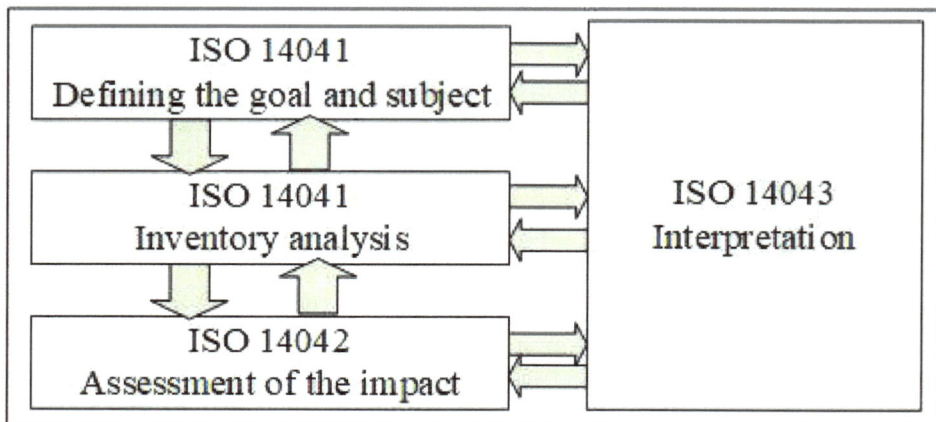

Fig. (4). LCA framework according to ISO 14040 [28].

Defining the goal and subject - in this phase, in addition to the goal and subject, the boundaries of the system (spatial and temporal) are set and the functional unit is defined. Defining a functional unit is an important basis for the normalization of input and output data in order to compare the results of the LCA study.

Analysis of Life Cycle Inventory (LCI) - refers to the collection, calculation and analysis of product system inputs and outputs. In this phase, all material and energy inputs and outputs are determined throughout the entire life cycle of the product or service. In LCA studies, data are mostly used from literature sources or based on databases such as Ecoinvent, EASEWASTE, SimaPro, GaBi, *etc.* [30].

Evaluation of Life Cycle Impact Assessment (LCIA) involves classifying LCI inputs and outputs into specific categories and inputs and outputs for each category based on indicators. The selection of impact categories is carried out on the basis of the set goal and subject, and there are appropriate indicators for each impact category. There are several LCIA methods such as EDIP, ReCiPe, USEtox, IPCC, *etc*. The impact categories are divided into two groups: standard impact categories and human health impact categories. The following categories are classified as standard impact categories: global warming potential, acidification, eutrophication, nutrient enrichment, ecotoxicity, and photochemical ozone formation. The following categories are classified into the categories of effects on human health: toxicity to humans (*via* water, air, and soil), toxicity of carcinogens and non-carcinogens to humans.

Thus, for example, global warming is a category of impact, and the indicator is the cumulative value of greenhouse gases released, which are expressed in terms of CO_2 equivalent. The Intergovernmental Panel on Climate Change (IPCC) has defined a model based on which the global warming potential is calculated *via* CO_2 equivalent. Gases CO_2, CH_4 and N_2O are gases that contribute to global warming, but in different ratios 1 CO_2 = 25 CH_4 = 298 N_2O [31].

Interpretation of life cycle results - is a procedure for identifying, verifying, qualifying and evaluating information obtained based on LCI and LCIA results. The results and their interpretation must be in accordance with the defined goal and subject.

Although the LCA methodology has been developed and harmonized with the ISO standard [28], the methodology is further complemented by guidelines [32]and numerous literature [33 - 35]. A comprehensive scope of the LCA is necessary to avoid shifting problems, for example, from one phase of the life cycle to another, from one region to another, or from one environmental problem to another [36].

LCA has found application in various fields such as agriculture [37, 38], water management [39 - 41], construction [42 - 44], food industry [45, 46], car industry [47, 48], and electronic equipment [49, 50].

Design for Environment

"Design for Environmental (DfE) is one of the possible approaches in the new product development process. The concept of DfE can be summarized in the principle 'do more, using less' which is the incentive for reducing the quantity of energy and material used to provide goods and services" [51].

DfE or Green or Eco Design is a new product design process, which in addition to the criteria that determine good product design, considers the impact of the product and all its stages of development on the environment. The purpose of this analytical tool is to assess and identify all the possibilities that this product induces in the environment throughout the entire life cycle [52].

Six principles of DfE are [53, 54]:

- Ensure sustainability of resources (This principle aims to address resource depletion by encouraging the reuse of resources within the techno-sphere, such as materials and components, and renewability of consumed resources, such as energy).

- Ensure healthy inputs and outputs (providing healthy inputs and outputs that will not affect environmental degradation and endanger human health; this principle requires the elimination of harmful substances and pollutants as well as the conversion of waste into useful materials for products and ecosystem),
- Ensure minimal use of resources in the production and transportation phase (encouraging designers to think about how product properties affect the efficiency of seemingly unrelated processes),
- Ensure minimal use of resources during use (this principle motivates product designers to be as efficient as possible in the use of energy and materials and their interaction with the user during the use phase in the product life cycle),
- Ensure appropriate durability of the product and components(this principle requires addressing two important strategies: long-term durability, along with the ability to update or upgrade products to current best practices).
- Enable disassembly, separation and purification (recycling, processing, reuse, repair and upgrading can be facilitated by incorporating these features through disassembly, separation and purification).

Examples of DfE are the use of old paper for the production of Figs. (**5** and **6**).

Fig. (5). Example of floor tiles and furniture made from old paper [55].

- Furniture, lighting and floor and wall tiles are made of old and recycled paper. In the world, more and more designers are working on the production of furniture, lighting and floor and wall tiles from old and recycled paper.
- Materials in construction or for the production of paper pellets, *i.e.* for the production of fuel. Newspapers and mixed paper can be used to make building materials to make plaster walls and various insulations.

Another example of the application of DfE is the creation of battery-free remote control for automobiles. This was done by utilizing the piezo effect, the charge created when crystals such as quartz are compressed. The remote is designed with a button on top and a flexible bottom. When the user pushes the button, the top button and flexible bottom compress the crystal, creating an electrical charge that powers a circuit to unlock the car. This device reduces the use of batteries and thus protects the environment [57].

Fig. (6). Demonstration of making wall coverings and pellets from old paper [56].

Ecological Economics

Ecological economics is a new transdisciplinary field of study that addressestherelationship between ecosystems and economic systems in the broadest sense. Theserelationships are central to many of humanity's current problems and to building a sustainable future but are not well covered by any existing scientific discipline [58]. Ecological economics is interested in the interactions between human behavior and the environment as a social-ecological system [59].

The conventional economy is primarily concerned with an isolated cycle of exchange between producers and consumers, without considering the depletion of natural resources, population growth and pollution. Ecological economics has taken a much broader view. And instead of focusing on the circular flow from producer to consumer, ecological economics recognizes that human economics is a smaller subsystem of the finite material system [58].

The ecosystem is a whole, and ecological economics is only a part of that system. The economy is based on the input of energy (solar energy) and natural resources (materials) which are transformed into products by various technological processes (Fig. **7**). In addition to products, residues and waste are generated in production systems. A significant element in the ecosystem is occupied by living organisms, *i.e.* their metabolism. Waste must be returned or is absorbed by the environment or processed by living organisms [60].

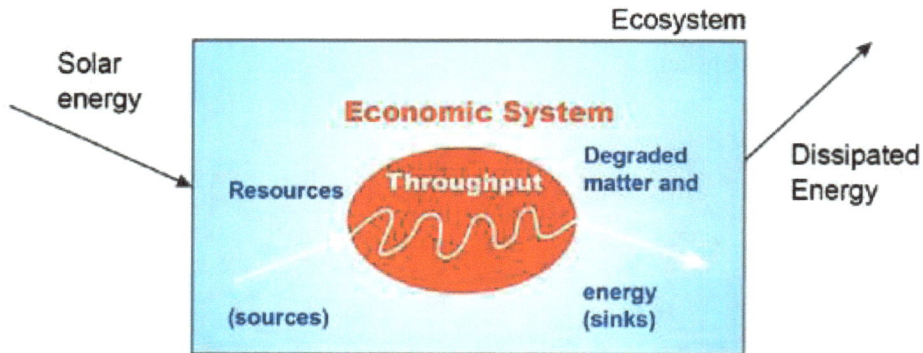

Fig. (7). The economy as an open system inside the ecosystem [60].

Ecological economics strives to place economic activities in the context of biological and physical systems, which reflect the way of life. The basic principle of ecological economics is that human activity must be limited by the ability of the environment to assimilate the loads produced by man. In short, the goal of the ecological economics is to live as long as we can with the limited resources available to us as well as a fairer distribution of resources.

Blue Economy

In the recent decade, the "Blue Economy" or "Oceans/Marine Economy" has been widely advocated by an array of interested partiesas a concept or a strategy for safeguarding the world's oceans and water resources. The concept of Blue Economy is originated from the United Nations Conference on Sustainable Development held in Rio de Janeiro in 2012 [61]. The United Nations offered a general definition of the "Blue Economy" as an ocean economy that aims at "the improvement of human well-being and social equity, while significantly reducing environmental risks and ecological scarcities [62]". More recently, the World Bank defined the "Blue Economy" as "the sustainable use of ocean resources for economic growth, improved livelihoods, and jobs while preserving the health of the ocean ecosystem [63]."

The blue economy is the sustainable use of ocean resources for economic growth, improved life expectancy and jobs while preserving the health of the ocean ecosystem [63].The blue economy is defined as all economic activities related to the oceans, seas and coasts. It covers a wide range of interconnected established and new sectors [64].The blue economy goes a step further where, by offering business innovative methods, it tries to prove that environmental sustainability and economic profitability do not have to be mutually exclusive.

In recent decades, human society has been completely separated from nature, that is, natural processes. All processes in nature function according to the laws of physics and chemistry, without the need for energy as well as without waste, *i.e.* the product of one process is always a raw material for another process. Everything functions in one circular process. On the other hand, man's actions through injustices in society, distorted exploitative economies have caused the disruption of natural ecosystems [65].In response to these problems, there is a demand for a new economy that will be more efficient and willing to cooperate.

Pauli distinguishes the red economy, that is, the one that only borrows and returns nothing, but only postpones for the next generations that will have to deal with a series of problems and lack of resources [66].The red economy focuses on one activity and forces one product. This product must be produced at low production prices while reducing labor costs. Such an economy takes only from nature, causes social and environmental problems, and on the other hand, poverty and unemployment are guaranteed by this approach. The green economy emerged in response to the red economy. The green economy has promoted green technologies, renewable energy sources, alternative production methods and materials Table **1**. Only one advantage has been put in focus and the effects on entire ecosystems have been ignored [67]. For example, palm oil used to make biodegradable soaps has led to the destruction of rainforests in Indonesia, or the production of shiitake mushrooms that is especially used in a vegetarian diet as a substitute for animal proteins has caused the cutting of Chinese oak trees to grow these mushrooms. This is how we got a problem solving in one place and problem creation in another. Green products, on the other hand, required large subsidies which is not good for the economy in the future.

Table 1. Tabular comparison of the red, green and blue economy [66].

-	Red Economy	Green Economy	Blue Economy
Production	High	Optimal	Optimal
Production costs	High	High	Low
Social capital	Low	Low	High
Waste	High	Low	No waste
Environmental damage	High	Low	No damage
Subsidy	High	High	Low
Investment	High	High	Low

The leading idea of Pauli's "Blue Economy" is the concept of cascading nutrients and energy.Thecascades of nutrients and energy mimic and function on the application of natural laws. Nutrients are broken down by microorganisms and bacteria, then plants take over those nutrients for growth and development, animals feed on plants, animals die and are broken down and become food for plants. The waste of one living being becomes food for another. This principle of the cascade of matter and energy is the basic starting point of the blue economy. Finding hundreds of the best technologies that are based on natural processes could affect the world economy and the sustainable provision of all human needs [67].

The Blue Economy as a project has begun to find the 100-best nature-inspired innovations that could impact the economy and provide basic human needs in a sustainable way - drinking water, food, work, energy and home (Pauli, 2010).

The main challengesfor the achievement of a true sustainable, Blue Economy are detailed as follows [68]:

- Sustainable use of biodiversity,
- Food security, focusing on the development of sustainable fish eriesor exploitation of wild fish stocks and sustainable and efficient aquaculture industries;
- Climate change and carbon budgets; facilitating the transition towards a low carbon economy and a renewable "blue" energy generation to address the acidification of oceans and pH decrease(CO_2cycle);and enhance blue carbon cycles or carbon sequestration cycles, linked to the damage of coastal habitats such as mangroves, seagrass meadows or salt marshes;
- Marine and coastal tourism, which continue showing growing patterns despite the global crisis. Increases of greenhouse gas emissions,water demand, sewage, waste generation, loss and degradation of coastal habitat, biodiversity and ecosystem services need to be addressed.
- Pollution and marine debris: a growing human population, the intensification of agriculture and urbanization of coastal areas, is at the land-origin of increasing marine pollution while shipping and marine resource exploitation (hydrocarbon or mining) are sea-based pollution sources.

Application of the Blue Economy

The Belgrade company "Eko fungi" (Serbia) is an example of the application of the blue economy in the Balkans. In 2013, this company started the industrial production of oyster mushrooms using a new technology that uses waste in the production of substrates, *i.e.* nutrient media, for growing mushrooms. This is the first plant of this type in Serbia and the world. Production of edible mushroom species such as oyster mushrooms, shiitake, *etc.* is traditionally based on the use of straw as a raw material for the production of substrates, which has started to become inaccessible and whose costs of transport and consumption of electricity in production have increased significantly.

This substrate production company uses any cellulosic material that can be procured within a radius of 30 km. So far, nowhere in the world has the project of production of substrates for growing edible plants on the basis of waste been realized. In addition, by applying the microbiological method - instead of energy, the power of microorganisms is used for heating thus eliminating the use of electricity. This approach reduces the use of electricity by as much as 90%. The spent substrate can be used in livestock nutrition or as an organic fertilizer, thus eliminating the generation of waste in the production process. In addition, this production can be applied on small farms throughout Serbia and employ a large number of households. The production capacity according to this technology is about 30t of edible mushrooms annually, *i.e.* processing of about 50t of cellulose waste [69].

Biomimicry

Biomimicry (bios - life and mimesis - imitate) represents an adaptive way of solving the problems of living and problems in the economy that can be found in nature, *i.e.* by imitating natural phenomena and processes. Biomimicry can still be defined as a set of innovations inspired by nature. Nature has always been an inexhaustible source of inspiration for scientists, experts, and artists from a variety of fields [70].

In nature, every organism is unique, and it is completely adapted to its environment and life and survival in all conditions. Observing a tree that emerges from a seed, which used nutrients and solar energy during its growth and development, it eventually dies and becomes food for a new life and beginning. Through evolution, nature has "solved" a large number of problems, *i.e.* plants, animals and microorganisms function in one harmony *i.e.* balance [71]. By studying organisms, that is, their adaptations and processes that take place in each organism individually, as well as interaction with other organisms and the environment, knowledge and new ideas for application in various fields of human creation such as industrial design, architecture, medicine, robotics, *etc.*, are gained.

Biomimicry is determined by three key principles [72]:

• Nature as a model - where by studying models from nature, these models are

evaluated by mimicry into new forms, processes, systems or strategies for solving technical problems,
- Nature as a measure - serves biomimicry to use environmental standards to assess the sustainability of innovation,
- Nature as a mentor - serves biomimicry to observe and perceive nature as a source of knowledge.

Biomimicry is advancing more and more with the development of technology, the synthesis of bio-sustainable and environmentally friendly materials, and the achievement of new scientific knowledge and achievements. The principles of biomimicry are based on the laws of nature [73]:

- Nature acts under the influence of sunlight,
- Nature uses only as much energy as it needs,
- Nature "recycles" everything,
- Nature acts on the principle of diversity,
- Nature rests on diversity,
- Nature adapts the form of functionality,
- Nature does not allow wastage.

The application of biomimicry in architecture and design is revolutionary. Man adapts to nature, improves his existence and survival in difficult conditions such as harsh climate, poor geological characteristics of the soil, difficult to access terrains, *etc.* By designing and realizing objects on the principle of biomimicry, it is easier for a person to overcome natural disasters such as earthquakes, floods, strong winds, increased radiation due to damage to the ozone layer. New ideas and innovative solutions in architecture strive to provide healthy working and living conditions in buildings of sustainable design and environmental efficiency [70, 74].

Application of Biomimicry in Practice

German botanist Barthlott, studying the lotus plant, noticed that the leaves are perfectly clean, that is, that impurities do not remain on the leaf. At first glance, the lotus leaves are smooth, but analyzing the surface, it was found that the surface is rough. The water that falls on the leaf, in the form of drops, slides from the leaves with impurities. The microscopic and nanoscopic architecture of the surface, which reduces the adhesion of dirt particles, is responsible for this effect on the surface of the lotus leaves. The surface of the lotus leaf is superhydrophobic, repelling water. The water that falls on the leaf forms balls, *i.e.* it does not adhere to the lotus surface [75]. Other plants, such as nasturtium

(*Tropaeolum sp.*), reed (*Phragmites sp.*), cabbage (*Brassica oleracea*), or columbine (*Aquilegia sp.*) show this effect, as do some animals (many insect wings).

The technical and economic importance of self-cleaning surfaces is increasing. This desired property of micro and nanostructuring of superhydrophobic surfaces is a purely physicochemical phenomenon and can be bionically transferred to technical surfaces. Meanwhile, with this product alone, there are about 600,000 buildings worldwide that are equipped with lotus effect surfaces.

The self-cleaning ability of water-repellent nanostructured surfaces was discovered in the 1970s by Wilhelm Barthlott. With regard to technical teaching for the application of the self-cleaning effect, Barthlott applied for international patent protection. Furthermore, products based on the technical instruction developed by Barthlott to apply the self-cleaning effect are internationally comprehensively protected by the trademarks "Lotus Effect". The exclusive owner of the brand is Sto AG from Stühlingen, a manufacturer of, among other things, facade paints "Lotusan", which Sto AG presented in 1999 as the first commercial product in the implementation of Barthlott's teaching on the market.

Another option is to make self-cleaning awnings, tarpaulins and sails on the principle of the lotus effect, which otherwise quickly become dirty and difficult to clean. Then making hydrophobic glass surfaces. This phenomenon of leaf cleaning has been used to create facade pains that are based on hydrophobicity and a special microstructure in order to achieve less adhesion of dirt and impurities, and to protect the facade from microclimatic influences. Initially, the aim was to make the facades as smooth and even as possible [76]. The Lotus building was made in China, which is currently a pearl of modern architecture. It is a low-energy facility, with 2,500 geothermal pipes conducted through an artificial lake.

Cradle to Cradle Design

There is no concept of waste in nature, that is, everything is food for other organisms or systems. The materials are reused in cycles, and there are no durable and bioaccumulative materials. And biota is created on the planet by the action of solar energy.

The production systems of the industrial revolution are based on a direct water-linear flow of material from the cradle to the grave - a model that takes, makes and consumes. Materials are taken from nature and refined, products are assembled, distributed, used by consumers, and then disposed of in landfills or incinerators. Each step in this course usually creates unintended impacts on the environment and health. The emergenceof modern industrial processes has had the added effect of making many processes and materials more toxic. Today, they are the legacy of a linear model that deeply affects the air we breathe, the water we drink, the climate in which we live, the diseases we suffer from, and global politics [77, 78].

In response to widespread environmental degradation, governments and industries have adopted a strategy known as "environmental efficiency" - minimizing waste, pollution and depleting natural resources. Many companies have achieved significant cost savings and reduced environmental impact by embracing environmental efficiency. But long-term prosperity does not depend on the efficiency of a fundamentally destructive linear model of the economy. This depends on the efficiency of processes that are designed to be sustainable, healthy and renewable [79].

Cradle to Cradle (C2C) Design offers an alternative. It rejects the assumption that human industry inevitably destroys the natural world, or that the demand for goods and services is the main cause of environmental diseases. Instead, it encompasses abundance, human ingenuity, and positive aspirations. Today, thanks to the growing knowledge of living Earth, our designs can reflect a new spirit. C2C Design incorporates this new environmental awareness at all levels of human endeavor. Its principles are built on the intelligence, abundance and efficiency of natural systems - the flows of energy and nutrients that support the Earth's biodiversity [80].

Natural ecosystems functionon some key principles that human design can emulate. First, there is no "waste" in nature; waste from one organism provides nutrients for another. Second, all life on Earth is stimulated by solar energy. Third, life thrives in diversity, constantly adapting to fill niches. C2C Design models the human industry on these natural principles. It envisions a sun-powered world in which growth is good, nutrient waste, and productive diversity enrich human and natural communities [80, 81].

The application of the C2C principle in the industry creates cyclical material flows (Cradle to Cradle, not Cradle to Grave) which, like the cycle of nutrients in the soil, eliminate the concept of waste. Each material in the product is designed to be safe and efficient and to provide high-quality resources for future generations of products. All materials are designed as nutrients, circulating safely and productively in one of two "metabolisms" [82]:

• Biological metabolism and
• Technical metabolism.

Biological metabolism is a system of natural processes that support life. These processes involve the degradation of organic matter and their incorporation into organisms - cyclic, and eventually stimulated by sunlight. Materials that contribute to the productivity of metabolism are biological nutrients. Ideally, human industry products designed from biodegradable, environmentally friendly materials participate in biological metabolism after use by decomposition.

C2C Design is a thinking system based on the belief that human design can approach more effectively by learning from nature and fitting its patterns. The industry can be transformed into a sustainable enterprise - one that creates economic, environmental and social value - with a thoughtful design that reflects the safe, regenerative productivity of nature and eliminates the concept of waste. C2C design is the key to creating a regenerative economy, inspired by natural systems, that benefits society, the economy and the environment. C2C is based on the following principles [83, 84]:

1. Waste is food. The first principle requires the elimination of the very concept of waste and encourages it to be inspired by endless cycles of nutrients in nature. Instead of an environmentally efficient approach to trying to reduce waste, the focus should be on designing systems with products that can be taken as nutrients through other processes. This also applies to emissions during the product production phase, as well as to the product itself after it reaches the disposal phase. To ensure that such emissions can be subjected to a 100% closed loop, recyclable materials should be defined as technical or biological nutrients. Technical nutrients should be designed for industrial recycling, while biological nutrients must be designed to return to the soil and feed the environment. Biological and technical nutrients should not be mixed outside of easy separation. Otherwise, a product is created that does not fit into either biological or technical "metabolism". Such a product can never really be recycled, but only recycled into a product of lower quality and value [79, 83].

2. Use of renewable energy sources. The second principle dictates that the energy needed to promote the Cradle closed circle into the Cradle society must come from what is called "current solar income", defined as photovoltaic, geothermal, windy, hydro and biomass. These sources correspond to the general understanding of renewable energy sources. Due to the vision to be fully supplied with energy from the sun, the design between the bed and the cradle is not limited by any restrictions in the use of energy during the life cycle of the product. As long as energy quality meets the requirements (current solar income) the amount of energy is irrelevant [79, 83].

3. Encouraging diversity. The main point of this last principle is to avoid solutions

for all sizes and instead design products and systems that take into account the local environment, economy and culture. They are also encouraged to "become native" and play their role as species among other species. Therefore, the goal should not be to reduce the negative impact on the environment as suggested by the concept of environmental efficiency, as this would result in isolation from other species [83].

Application of C2C

Upholstery fabrics, which wear out with use, can be designed from biological nutrients that can be returned to ecosystems after use. Climatex® Lifecycle™ fabric is an example of this type of product. The fabric is a mixture of wool without pesticide residues and organically grown ramie (a plant from the nettle family), dyed and completely treated with non-toxic chemicals. All product and process intakes are defined and selected for their human and environmental safety in the context of biological metabolism. Currently, fabric liners (process "waste") are used by garden clubs as mulch for growing fruits and vegetables, thus returning to textile biological nutrientsfood for new growth. The return of nutrients from the life cycle of Climatex to biological metabolism depends on its application and handling after use. The design must facilitate the clean separation of the fabric from materials that cannot function as biological nutrients, such as synthetic foams. Substances used for fabric treatment and cleaning should also be compatible with biological metabolism. Finally, the nutrient cycle of fabric use is likely to rely on a well-managed composting system [82].

The technical metabolism industry can also be modeled by natural processes to create technical metabolisms, systems that produce industrial materials productively. These materials, valuable in terms of performance and usually "non-renewable", are technical nutrients designed to circulate safely and continuously through product life cycles from C2C production, use, recovery and re-production. The processes of technical metabolism are industrial, not natural, but they are ideally stimulated by solar energy as well, in the form of renewable energy.

A well-developed lead-acid battery recovery system provides a provocative model for the development of technical metabolisms. Car batteries are useful to customers for storage and provision of electricity by design, but along the way they pose a risk to customers and the environment as a result of the hazardous substances they contain. In order to reduce the risk of releasing hazardous substances, economic incentives have been built in to encourage the return of old batteries to authorized locations, with the use of a new battery. Old batteries are sent to secondary smelters where the material value of lead, plastic and acid is collected for use in new batteries. Over 95% of all lead and plastic from remanufactured car batteries is recycled, making it the most recycled consumer product in the United States.

Technical metabolisms exist in the natural world and the release of materials into ecosystems is inevitable, and technical nutrients should ideally pose little or no danger to biological metabolism. Lead is universally recognized as so toxic that even minor release harms human and environmental health. For carbatteries, there are safer alternatives (*e.g.*, lithium, zinc) that provide comparable performance. Although these alternatives are currently more expensive, they do not carry the associated environmental and health costs of lead that we all currently subsidize. The C2C principle suggests replacing lead, not just to optimize the technical metabolism of lead-acid batteries to reduce human and environmental impact.

The C2C concept is a new concept of modeling the economy through processes similar to nature where materials are viewed as nutrients circulating within biological or technical metabolism into a healthy,*i.e.* safe metabolism. This healthy metabolism allows for reuse, material optimization, recycling and nature conservation. The C2C concept shapes materials and objects in the image of natural processes. This new concept sees artificially created materials, obtained in other industrial processes, as nutrients that circulate endlessly in the healthy metabolism of the living world. In this way, an infinite life cycle of inanimate matter is created in the image of the living world, where the biosphere is equated with the technosphere.

The Swiss company Rohner uses 16 colors for its fabrics that do not contain any toxic substances. By mixing these colors, any shade can be obtained. Biodegradable felt is made from the waste that remains in the production process during cutting. This felt is used in agriculture in the winter and is completely decomposed by spring, creating nutrients [85].

Since 2017, C & A has started phasing out clothing models made from non-renewable materials, with a focus on biological materials that can be reused, recycled and eventually safely decomposed [86].

The concept of the circular economy is increasingly used in construction, and in this branch innovations in the use and revitalization of old materials and their re-incorporation into some new formare increasingly being introduced. Urban agriculture is increasingly being realized within self-sustaining urban areas. Fruits and vegetables are grown on the roofs of buildings, that is, food is grown in cities without transporting food from other areas in containers.

Another example of the application of the C2C concept in construction is the reconstruction of the roof of a Ford car factory in the USA (Fig. **8**). Ford opted to build a green roof instead of building a water treatment plant. The green roof was built on an area of 40,000 m², where the grass on the roof of this industrial complex collects, purifies atmospheric water [87].

Fig. (8). Appearance of the roof of the ford factory, USA [88].

The architectural firmSPARK in Singapore is another example of the application of the C2C concept through the construction of vertical aquaponic agriculture. In this building, a system of circulating waste nutrients from the bottom of the building to the top, processing of waste materials to obtain heat and electricity, as well as savings by maintaining a modern lighting system, has been established. The system is completely self-sustaining. In addition, this building houses elderly people in several residential units. These people work in the garden, socialize and educate about growing plants in urban vertical agriculture which has an important social component.

Cleaner Production

Cleaner production is the continuous application of an integrated, preventive environmental strategy to processes, products and services to increase eco-efficiency and reduce risks to humans and the environment (Fig. **9**) [89].

Cleaner productions focus on conserving water, energy and natural materials. This is a constant review of products, processes and services in order to strive for the principles of sustainable development [90]. Cleaner production is a general term that describes a preventive environmental approach, aimed at increasing resource efficiency and reducing the generation of pollution and waste at source, rather than addressing and mitigating just the symptoms by technically "treating" an existing waste/pollution problem. Cleaner production addresses the problem at several levels at once, serving as a holistic integrated preventive approach to environmental protection. In other words, Cleaner production avoids the end-of-pipe approach [91].

Cleaner production focuses on minimizing resource use and avoiding the creation of pollutants, instead of the current practice of trying to manage pollutants once they have been generated. This approach is implemented by reviewing products, processes and services with a view to sustainable development. The basic CP estimation procedure is shown in the figure. The goal of the process is to identify waste streams and provide a plan to reduce energy and material loss [91].

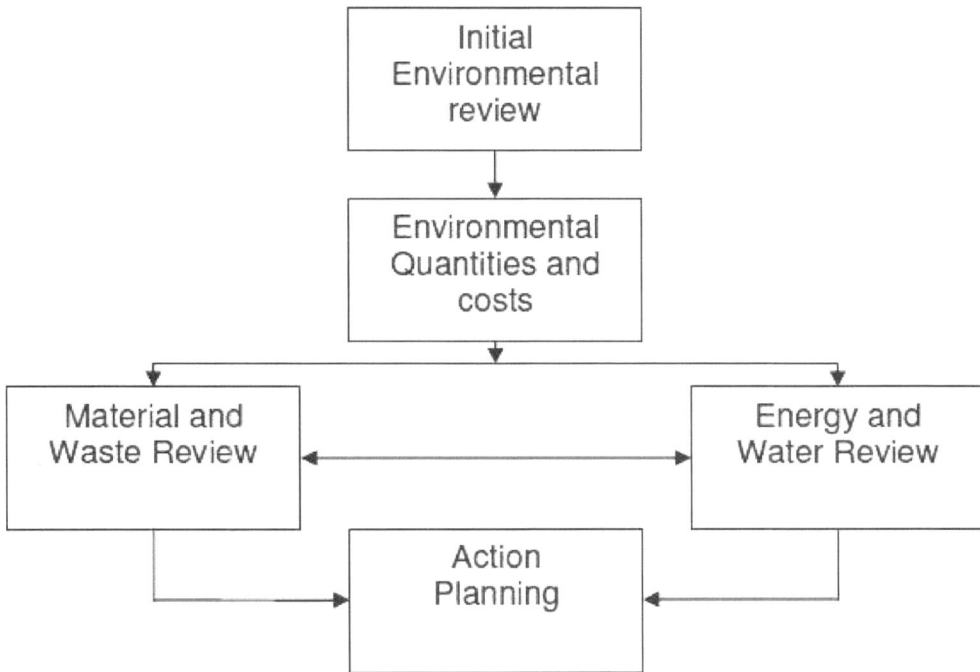

Fig. (9). Basic principles of cleaner production [91].

Cleaner production is based on the following principles [92]:

• Precaution principle - which requires the person proposing and establishing the process to prove that the process or activity will not cause harm, *i.e.* that it will be safe and effective. Thus, the burden of proof was transferred from society to the field of "polluters". This principle is an integral part of sustainable development. This principle is also called the reduction of anthropogenic inputs to the environment.

• Principle of prevention - which emphasizes that it is far easier and more efficient to prevent damage than to repair it later. This principle requires control and supervision of the process from beginning to end, instead of the current control at the end of the process. The preventive nature of cleaner production requires a new approach to reviewing product design, consumer demand, material consumption patterns. Society is required to adopt an integrated approach to resource use and exploitation. This integrated approach ensures that the problem is solved without causing another pollution problem.

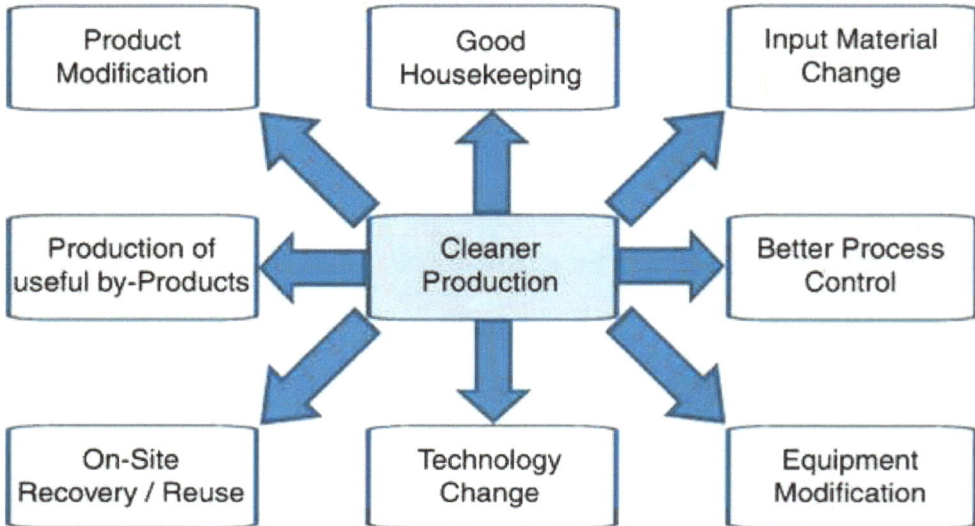

Fig. (10). General goals of cleaner production [93].

The main directions of cleaner production are shown in Fig. **(10)** [92, 93].

- Good housekeeping - taking appropriate management and operational conditions to prevent leaks and spills (such as a preventive maintenance plan and frequent equipment inspections) and implementing existing work instructions.
- Input substitution - replacement of input materials with less toxic or renewable materials or additional materials that have a longer service life in production.
- Better process control - modification of operating procedures, machine instructions and process logging in order to increase process efficiency and lower waste and emission rates.
- Equipment modification - modification of existing production equipment and utilities, for example, by adding measuring and control devices, in order to process them with greater efficiency and lower waste and emission rates.
- Technology change –change of technology, treatment sequence and/or synthesis path to reduce waste generation and emissions during production.
- Product modification - modification of product characteristics in order to minimize the impact of the product on the environment during or after its use

(disposal) or to minimize the impact of its production on the environment.

- Efficient use of energy - energy is a very significant source of environmental impact. Energy use can have an impact on land, water, air, biodiversity, as well as on the production of large amounts of solid waste. The environmental impact of energy use can be reduced by improving energy efficiency and using energy from renewable sources, such as the sun and wind.
- On-site recovery/reuse - reuse of waste materials in the same process or for some other useful use in the company.

Cleaner production can reduce operating costs, improve profitability and protect the health of employees, as well as reduce the negative impact of production on the environment. Companies are often surprised by cost reductions through the adoption of cleaner production techniques. Often, minimal or no capital expenditures are required to achieve useful gains. Obviously, cleaner production techniques are a good deal for the industry because they will [94]:

- Reduce waste disposal costs,
- Reduce raw material costs,
- Reduce costs for ensuring a healthy environment,
- Improve public relations, *i.e.* the reputation of the company,
- Improve the performance of the company,
- Improve competitiveness in the domestic and international market.

Application of Cleaner Production

Danish fish processing industry has been one of the pioneers in the implementation of cleaner production and environmental management systems. As early as the end of 1980, the analysis of solutions in this industry on the principle of Cleaner production began. The implementation of the Cleaner production principle in a period of about 15 years has made significant strides in terms of reducing water consumption, wastewater emissions and the use of fish waste for valuable by-products. However, more attention can be paid in the future to reducing energy consumption, changing the types of packaging and the impact on the environment at other stages of the product life cycle. The Danish authorities and companies have mainly focused on reducing on-site wastewater emissions, but Life Cycle Assessments show that more attention needs to be paid to reducing environmental impacts in other parts of the product chain, such as fishing operations and transport [95].

Regenerative Design

The term regenerative means the process by which one's own sources of energy and materials are renewed or revitalized, creating sustainable systems that integrate the needs of society with the integrity of nature. The Regenerative design seeks to address the continued degradation of ecosystems by developing the built environment to restore the capacity of ecosystems to function at optimal health for the mutual benefit of both human and non-human lives [96].

The Regenerative design is a different approach to development. Instead of cutting down forests, draining swamps, or concreting green areas to make room for wooden, concrete or steel elements or buildings, an environment is built that integrates into the existing landscape. It is a sophisticated design approach that relies on state-of-the-art technologies and indigenous techniques to create thoughtful products and designs, which will not create a barrier between built and natural environments, through high-efficiency buildings "nested" in functional landscapes. Regenerative design strives for a multiple development paths that reconcile built and natural systems in one whole (Fig. **11**) [97].

Regenerative design is a design principle in which we view nature as our guide to everything we do. Long-term evolutionary processes in nature take only what it needs without consuming resources as well as without generating waste. We should use natural processes as a model for the "recovery" of our planet, by imitating biology to move in the direction of "returning" to the planet and not just the previous "consumption" [97].

For example, solar panels are built on the simple idea of mimicking the leaves on a tree, which absorb the sun's rays and convert them into energy. It is an example of biomimicry, *i.e.* inspiration from nature for solving life problems. On the other hand, wood can also be a role model for us. A tree takes in the sunlight, with water from the soil it takes in nutrients and produces organic matter (wood mass). Leaves that fall during the year, as well as branches, are slowly decomposed with the help of microorganisms and fungi and returned to the soil as a nutrient. This nutrientserves other plants and animals.It is a process that happens continuously, every day throughout the year. In this way, the tree returned to the soil the nutrients it took from the soil, it did not create waste that needs to be taken somewhere and still processed. If a man were to try to apply this example in his household or business, taking it from nature but also returning it to the earth, the so-called regenerative sustainability would be established.

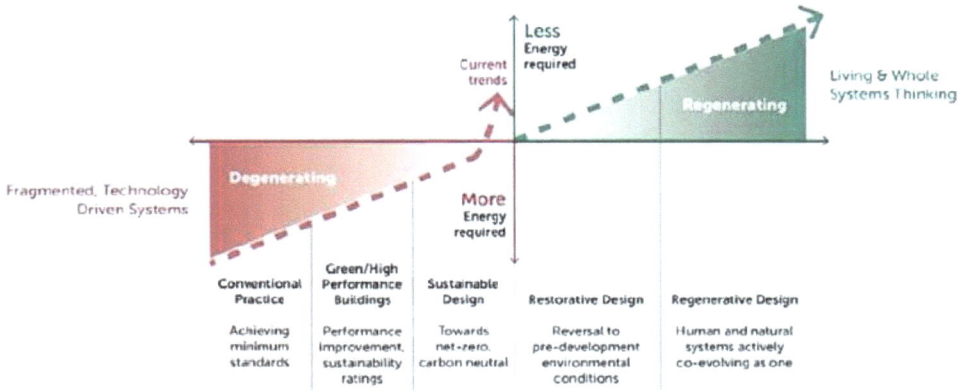

Fig. (11). Contrast of technical system design and living system design [98].

The figure shows the range and trajectory of different approaches to sustainability to put the emphasis on the need to transition from degenerating to regenerative systems. The fragmented, technologically impaired systems visible in today's conventional, green, and highly efficient buildings can only hope to achieve a net-zero or carbon effect. Existing approaches to sustainable design in the built environment generally fall on these degenerating systems, as there is a strong focus on reducing any negative design impacts by improving the efficiency of its components. In order to move to regenerative design, less emphasis needs to be placed on the insulated element or building, and more on the design process that is focused on the evolution of the entire system [98].

Regenerative sustainability is the idea of establishing a socio-ecological system in which people do not focus only on reducing and limiting their own negative impact on nature, but actively contribute to regeneration and positively affect the natural, social and ecological system [97].

Regenerative sustainability implies:

- Man is an integral part of nature,
- Consideration of problems and solutions from a systemic perspective and contextually,
- Involvement of the community in the discussion, development and implementation of solutions,
- Ecosystems and processes in nature provide inspiration for solution development.

An Example of Regenerative Design

VanDusen Botanical Garden in Vancouver was opened in 1975 and covers 22 hectares. The garden contains more than 250,000 plants from around the world (Africa, South America, the Himalayas, the Mediterranean, *etc.*). During 2011, the construction of the visitor center within the botanical garden was completed, in which $ 22 million was invested. The visitor center was designed and built on the principle of regenerative design, *i.e.* by establishing a balance between visual and ecological, as well as between architecture and landscape. The roof of the center is wavy and meanders like flower petals, and the central part of the building is the oculus so that the whole building fits in with the surrounding landscape (Fig. **12**) Within the center, there is a cafe, library, shop, offices, classrooms, *etc*. The visitor center makes a number of innovations and solutions such as:

• Geothermal wells, solar photovoltaic cells, solar hot water pipes,
• Construction material is wood,
• Rainwater is used as technical water,
• Bioreactor treatment of sanitary water.

With these innovative solutions, the reduction of harmful emissions and energy to zero on an annual level has been achieved.

Fig. (12). VanDusen Botanical Garden with a visitor center in Vancouver [99].

CONCLUDING REMARKS

This chapter provides a brief overview of the wide range of possibilities in improving production and the product itself, in accordance with the laws that function in nature. The current approach in production and the economy is based on the depletion of natural resources, the generation of large amounts of waste and the production of pollutants that have a negative impact on human health and the environment. This paper deals with a large number of possibilities starting from the circular economy, industrial economy, ecological blue economy, biomimicry, and cradle to cradle, cleaner production and regenerative design. Common to all the above possibilities is that new approaches are being considered in the production process, product design and the product itself through reducing waste generation, use of materials that are less harmful and dangerous to the environment, design of production and products that are based on natural processes or mimic natural processes. With this new approach in production, we strive to live in harmony with nature, where our way of life contributes to reducing the negative impact on the environment and excessive depletion of natural resources.

CONSENT FOR PUBLICATION

Not applicable.

CONFLICT OF INTEREST

The author declares no conflict of interest, financial or otherwise.

ACKNOWLEDGEMENT

Declared none.

REFERENCES

[1]   M. Črnjar, K. Črnjar, J. Perić, R. Zelenika, and N. Denona-Bogović, *Sustainable development management: economics, ecology, environmental protection.* Faculty of Management in Tourism and Hospitality, 2009.

[2]   M. Weber, and M. Weber, *Towards environmental innovation systems.,* J. Hemmelskamp, Ed., Springer: Berlin, 2005.
       [http://dx.doi.org/10.1007/b138889]

[3]   A. Diekmann, and P. Preisendörfer, *Umweltprobleme als Allmende-Dilemma. Dies. Umweltsoziologie. Eine Einführung.* Rowohlt: Reinbek, 2001.

[4]   T. Lahti, J. Wincent, and V. Parida, "A definition and theoretical review of the circular economy, value creation, and sustainable business models: where are we now and where should research move in the future?", *Sustainability (Basel),* vol. 10, no. 8, p. 2799, 2018.
       [http://dx.doi.org/10.3390/su10082799]

[5]   L. Kok, G. Wurpel, and A. Ten Wolde, "Unleashing the power of the circular economy", *Report by IMSA Amsterdam for Circle Economy.,* 2013.

[6]   J. Korhonen, "Theory of industrial ecology. Progress in Industrial Ecology", *Int. J.,* vol. 1, pp. 61-88, 2004.

[7]   M. Geissdoerfer, P. Savaget, N.M.P. Bocken, and E.J. Hultink, "The Circular Economy – A new

sustainability paradigm?", *J. Clean. Prod.,* vol. 143, pp. 757-768, 2017.
[http://dx.doi.org/10.1016/j.jclepro.2016.12.048]

[8] A. Heshmati, "A review of the circular economy and its implementation", *International Journal of Green Economics,* vol. 11, no. 3/4, pp. 251-288, 2017.
[http://dx.doi.org/10.1504/IJGE.2017.089856]

[9] M.E. Moula, J. Sorvari, and P. Oinas, *Constructing a green circular society,* 2017.
[http://dx.doi.org/10.31885/2018.00002]

[10] E. MacArthur, "Towards the circular economy", *J. Ind. Ecol.,* vol. 2, pp. 23-44, 2013.

[11] M. Drljača, "The concept of circular economy Quality & Excellence", *Foundation for Quality Culture and Excellence, Belgrade,* vol. 4, pp. 18-22, 2015.

[12] A. Bruel, J. Kronenberg, N. Troussier, and B. Guillaume, "Linking industrial ecology and ecological economics: A theoretical and empirical foundation for the circular economy", *J. Ind. Ecol.,* vol. 23, no. 1, pp. 12-21, 2019.
[http://dx.doi.org/10.1111/jiec.12745]

[13] P. Lacy, J. Long, and W. Spindler, *The Circular Economy Handbook: Realizing the Circular Advantage.* Springer Nature, 2020.
[http://dx.doi.org/10.1057/978-1-349-95968-6]

[14] F.S. Lyakurwa, "Industrial Ecology A New Path To Sustainability: An Empirical Review", *Indep. J. Manag. Prod.,* vol. 5, no. 3, pp. 623-635, 2014.
[http://dx.doi.org/10.14807/ijmp.v5i3.178]

[15] F. Hond, "Industrial ecology: a review", *Reg. Environ. Change,* vol. 1, no. 2, pp. 60-69, 2000.
[http://dx.doi.org/10.1007/PL00011534]

[16] J. Korhonen, A. Honkasalo, and J. Seppälä, "A.nHonkasalo, andJ. Seppälä, "Circular economy: the concept and its limitations"", *Ecol. Econ.,* vol. 143, pp. 37-46, 2018.
[http://dx.doi.org/10.1016/j.ecolecon.2017.06.041]

[17] B. Steubing, H. Böni, M. Schluep, U. Silva, and C. Ludwig, "Assessing computer waste generation in Chile using material flow analysis", *Waste Manag.,* vol. 30, no. 3, pp. 473-482, 2010.
[http://dx.doi.org/10.1016/j.wasman.2009.09.007] [PMID: 19793641]

[18] P.H. Brunner, "Materials flow analysis and the ultimate sink", *J. Ind. Ecol.,* vol. 8, no. 3, pp. 4-7, 2004.
[http://dx.doi.org/10.1162/1088198042442333]

[19] P. Baccini, and P.H. Brunner, *Metabolism of the anthroposphere: analysis, evaluation, design.* The MIT Press Cambridge: London, Massachusetts, England, 2012.

[20] P.H. Brunner, and H. Rechberger, *Practical Handbook of material flow analysis.* Lewis Publishers CRC press: Boca Raton/London/New York/Washington D.C/Florida, 2004.

[21] R. Fehringer, B. Brandt, P.H. Brunner, H. Daxbeck, S. Neumayer, and R. Smutny, ". Guidelines for the use of material flow analysis (MFA) for municipal solid waste (MSW) management. Project AWAST", In: *Aid in the management and European Comparison of municipal solid waste treatment methods for a global and sustainable approach.* Vienna University of Technology & Resource Management Agency, 2004.

[22] H. Stevanović-Čarapina, N. Žugić-Drakulić, A. Mihajlov, and I. Čarapina-Radovanović, "MFA and LCA as analytical instruments in the field of environment", *Environment towards Europe", Limes. Journal of Social Sciences and Humanities.,* vol. vol.1, 2014.

[23] P.H. Brunner, H. Daxbeck, and P. Baccini, "Industrial metabolism at the regional and local level: A case study on a Swiss region", In: *Industrial Metabolism—Restructuring for sustainable development.,* R.B. Ayres, U.E. Simonis, Eds., United Nations University Press: Tokyo, 1994.

[24] S. Li, Z. Yuan, J. Bi, and H. Wu, "Anthropogenic phosphorus flow analysis of Hefei City, China", *Sci. Total Environ.,* vol. 408, no. 23, pp. 5715-5722, 2010.

[http://dx.doi.org/10.1016/j.scitotenv.2010.08.052] [PMID: 20863550]

[25]   Z. Yuan, J. Shi, H. Wu, L. Zhang, and J. Bi, "Understanding the anthropogenic phosphorus pathway with substance flow analysis at the city level", *J. Environ. Manage.,* vol. 92, no. 8, pp. 2021-2028, 2011.
[http://dx.doi.org/10.1016/j.jenvman.2011.03.025] [PMID: 21489683]

[26]   S. Barles, "Urban metabolism of Paris and its region", *J. Ind. Ecol.,* vol. 13, no. 6, pp. 898-913, 2009.
[http://dx.doi.org/10.1111/j.1530-9290.2009.00169.x]

[27]   H.D. Stevanović-Čarapina, J.M. Stepanov, D.C. Savić, and A.N. Mihajlov, "Emission of toxic components as a factor in choosing the best option for waste management using the concept of Life Cycle Assessment", In: *Chemical industry.* vol. Vol. 65. Hemijskaindustrija, 2011, pp. 205-211.

[28]   H.D. Stevanović-Čarapina, J.M. Stepanov, D.C. Savić, and A.N. Mihajlov, "Emission of toxic components as a factor in choosing the best option for waste management using the concept of Life Cycle Assessment", In: *Chemical industry.* vol. Vol. 65. Hemijskaindustrija, 2011, pp. 205-211.

[29]   European Commission, *Report from the Commission to the European parliament, the council, the european economic and social committee and the committee of the regions, on the Thematic Strategy on the Prevention and Recycling of Waste,* 2011.

[30]   A. Laurent, J. Clavreul, A. Bernstad, I. Bakas, M. Niero, E. Gentil, T.H. Christensen, and M.Z. Hauschild, "Review of LCA studies of solid waste management systems – Part II: Methodological guidance for a better practice", *Waste Manag.,* vol. 34, no. 3, pp. 589-606, 2014.
[http://dx.doi.org/10.1016/j.wasman.2013.12.004] [PMID: 24388596]

[31]   H. Tian, X. Xu, C. Lu, M. Liu, W. Ren, and G. Chen, "Net exchanges of CO_2, CH_4, and N_2O between China's terrestrial ecosystems and the atmosphere and their contributions to global climate warming", *Journal of Geophysical Research: Biogeosciences.,* vol. vol.116, 2011.

[32]   J.B. Guinée, M. Gorrée, R. Heijungs, G. Huppes, R. Kleijn, A. de Koning, L. van Oers, A. Wegener Sleeswijk, S. Suh, H.A. Udo de Haes, J.A. de Brujin, R. van Duin, and M.A.J. Huijbregts, *Handbook on Life Cycle Assessment. Operational Guide to the ISO Standards.IIa: Guide. IIb. Operational annexes.* Kluwer Academic: Dordrecht, 2002, p. 692.

[33]   F.R. McDougall, P.R. White, M. Franke, and P. Hindle, *Integrated solid waste management: a life cycle inventory.* John Wiley & Sons, 2008.

[34]   H. StevanovićČarapina, *A.Jovović, and J. Stepanov, Life Cycle Assessment (LCA) as an instrument in strategic waste management planning.Monography.* Educons University, 2011.

[35]   H. Wenzel, M.Z. Hauschild, and L. Alting, *Environmental assessment of products.* vol. 1. Hingham MA, USA: Chapman & Hall, United Kingdom, Kluwer Academic Publishers, 1997.
[http://dx.doi.org/10.1007/978-1-4615-6367-9]

[36]   G. Finnveden, M.Z. Hauschild, T. Ekvall, J. Guinée, R. Heijungs, S. Hellweg, A. Koehler, D. Pennington, and S. Suh, "Recent developments in life cycle assessment", *J. Environ. Manage.,* vol. 91, no. 1, pp. 1-21, 2009.
[http://dx.doi.org/10.1016/j.jenvman.2009.06.018] [PMID: 19716647]

[37]   P. Goglio, W.N. Smith, B.B. Grant, R.L. Desjardins, B.G. McConkey, C.A. Campbell, and T. Nemecek, "Accounting for soil carbon changes in agricultural life cycle assessment (LCA): a review", *J. Clean. Prod.,* vol. 104, pp. 23-39, 2015.
[http://dx.doi.org/10.1016/j.jclepro.2015.05.040]

[38]   S. Keyes, P. Tyedmers, and K. Beazley, "Evaluating the environmental impacts of conventional and organic apple production in Nova Scotia, Canada, through life cycle assessment", *J. Clean. Prod.,* vol. 104, pp. 40-51, 2015.
[http://dx.doi.org/10.1016/j.jclepro.2015.05.037]

[39]   A. Ahmadi, and L. Tiruta-Barna, "A Process Modelling-Life Cycle Assessment-MultiObjective Optimization tool for the eco-design of conventional treatment processes of potable water", *J. Clean.*

Prod., vol. 100, pp. 116-125, 2015.
[http://dx.doi.org/10.1016/j.jclepro.2015.03.045]

[40] G. Barjoveanu, I.M. Comandaru, G. Rodriguez-Garcia, A. Hospido, and C. Teodosiu, "Evaluation of water services system through LCA. A case study for Iasi City, Romania", *Int. J. Life Cycle Assess.,* vol. 19, no. 2, pp. 449-462, 2014.
[http://dx.doi.org/10.1007/s11367-013-0635-8]

[41] L. Wu, X.Q. Mao, and A. Zeng, "Carbon footprint accounting in support of city water supply infrastructure siting decision making: a case study in Ningbo, China", *J. Clean. Prod.,* vol. 103, pp. 737-746, 2015.
[http://dx.doi.org/10.1016/j.jclepro.2015.01.060]

[42] L. Huang, R.A. Bohne, A. Bruland, P.D. Jakobsen, and J. Lohne, "Life cycle assessment of Norwegian road tunnel", *Int. J. Life Cycle Assess.,* vol. 20, no. 2, pp. 174-184, 2015.
[http://dx.doi.org/10.1007/s11367-014-0823-1]

[43] B. Pang, P. Yang, Y. Wang, A. Kendall, H. Xie, and Y. Zhang, "Life cycle environmental impact assessment of a bridge with different strengthening schemes", *Int. J. Life Cycle Assess.,* vol. 20, no. 9, pp. 1300-1311, 2015.
[http://dx.doi.org/10.1007/s11367-015-0936-1]

[44] S.V. Russell-Smith, and M.D. Lepech, "Cradle-to-gate sustainable target value design: integrating life cycle assessment and construction management for buildings", *J. Clean. Prod.,* vol. 100, pp. 107-115, 2015.
[http://dx.doi.org/10.1016/j.jclepro.2015.03.044]

[45] A.K. Cerutti, S. Bruun, D. Donno, G.L. Beccaro, and G. Bounous, "Environmental sustainability of traditional foods: the case of ancient apple cultivars in Northern Italy assessed by multifunctional LCA", *J. Clean. Prod.,* vol. 52, pp. 245-252, 2013.
[http://dx.doi.org/10.1016/j.jclepro.2013.03.029]

[46] M. Tyszler, G. Kramer, and H. Blonk, "Comparing apples with oranges: on the functional equivalence of food products for comparative LCAs", *Int. J. Life Cycle Assess.,* vol. 19, no. 8, pp. 1482-1487, 2014.
[http://dx.doi.org/10.1007/s11367-014-0762-x]

[47] E.A. Nanaki, and C.J. Koroneos, "Comparative economic and environmental analysis of conventional, hybrid and electric vehicles – the case study of Greece", *J. Clean. Prod.,* vol. 53, pp. 261-266, 2013.
[http://dx.doi.org/10.1016/j.jclepro.2013.04.010]

[48] W-P. Schmidt, E. Dahlqvist, M. Finkbeiner, S. Krinke, S. Lazzari, D. Oschmann, S. Pichon, and C. Thiel, W.P, "Life cycle assessment of lightweight and end-of-life scenarios for generic compact class passenger vehicles", *Int. J. Life Cycle Assess.,* vol. 9, no. 6, pp. 405-416, 2004.
[http://dx.doi.org/10.1007/BF02979084]

[49] L. Deng, C.W. Babbitt, and E.D. Williams, "Economic-balance hybrid LCA extended with uncertainty analysis: case study of a laptop computer", *J. Clean. Prod.,* vol. 19, no. 11, pp. 1198-1206, 2011.
[http://dx.doi.org/10.1016/j.jclepro.2011.03.004]

[50] D. Elduque, C. Javierre, C. Pina, E. Martínez, and E. Jiménez, "Life cycle assessment of a domestic induction hob: electronic boards", *J. Clean. Prod.,* vol. 76, pp. 74-84, 2014.
[http://dx.doi.org/10.1016/j.jclepro.2014.04.009]

[51] *M. Bevilacqua, F.E.Ciarapica, and G.Giacchetta, Design for Environment as a Tool for the Development of a Sustainable Supply Chain.* Springer Science & Business Media, 2012.

[52] H. Stevanović-Čarapina, and A. Mihajlov, *Methodologies For Green Products Design.,* 2010.

[53] M.S.J. Hossain, and M. Iqbal, *The Trends of Environment Friendly Product Design in Bangladesh,* 2012.

[54] C. Telenko, C.C. Seepersad, and M.E. Webber, "A compilation of design for environment principles

and guidelines.InASME 2008 International Design Engineering Technical Conferences and Computers and Information in Engineering Conference", In: *American Society of Mechanical Engineers Digital Collection.*, 2008.
[http://dx.doi.org/10.1115/DETC2008-49651]

[55]　https://www.bloglovin.com/blogs/design-milk-31264/paperscapes-from-recycled-paper-to-funtional-47850876

[56]　https://www.kenaf-fiber.com/en/isolcell.htmlhttp://www.pellet-making.com/application/paper-pellet-machine.html

[57]　D.P. Fitzgerald, J.W. Herrmann, P.A. Sandborn, L.C. Schmidt, and T.H. Gogoll, "Design for environment (DfE): strategies, practices, guidelines, methods, and tools", *Environmentally conscious mechanical design.*, vol. 1, pp. 1-24, 2007.

[58]　R. Costanza, H.E. Daly, and J.A. Bartholomew, "Goals, agenda, and policy recommendations for ecological economics", *Ecological economics: The science and management of sustainability*, p. 525, 1991.

[59]　S. Heckbert, T. Baynes, and A. Reeson, "Agent-based modeling in ecological economics", *Ann. N. Y. Acad. Sci.*, vol. 1185, no. 1, pp. 39-53, 2010.
[http://dx.doi.org/10.1111/j.1749-6632.2009.05286.x] [PMID: 20146761]

[60]　C. CavalCanti, "Conceptions of ecological economics: Its relationship with mainstream and environmental economics", *Estud. Av.*, vol. 24, pp. 53-67, 2010.

[61]　UNCTAD, "United Nations Conference on Trade and Development", *The Ocean Economy: Opportunities and Challenges for Small Island Developing States,available at,* 2014.http://unctad.org/en/publicationslibrary/ditcted2014d5_en.pdf

[62]　United Nations, *Blue Economy Concept Paper Available at ,* 2014.*Blue Economy Concept Paper,* 2014.https://sustainabledevelopment.un.org/concent/documents/2978BEconcept.pdf

[63]　The World Bank, *What is the Blue Economy?,* 2017.

[64]　European Union, *The 2018 Annual Economic Report on EU Blue Economy",* 2018.

[65]　J.C. Pereira, "Environmental issues and international relations, a new global (dis)order - the role of International Relations in promoting a concerted international system", *Rev. Bras. Polit. Int.*, vol. 58, no. 1, pp. 191-209, 2015.
[http://dx.doi.org/10.1590/0034-7329201500110]

[66]　G.A. Pauli, *The blue economy: 10 years, 100 innovations, 100 million jobs.* Paradigm publications, 2010.

[67]　G.A. Pauli, *Blue Economy Principles,* 2012.

[68]　Available from: https://www.google.com/url?sa=t&rct=j&q=&esrc=s&source=web&cd=&ved=2ah UKEwjt--6Uu6DuAhVRpIsKHQvqAYkQFjACegQIAhAC&url=http%3A%2F%2Fwww.fao. org%2F3%2Fca7454en%2FCA7454EN.pdf&usg=AOvVaw23H_EWndke3L63m6yaBtdt

[69]　https://www.aurea.rs/2017/02/06/eko-fungi-proizvodnja-jestivih-gljiva-na-bazi-celuloznih-ostataka/

[70]　E. Kennedy, D. Fecheyr-Lippens, B.K. Hsiung, P.H. Niewiarowski, and M. Kolodziej, "Biomimicry: A path to sustainable innovation", *Des. Issues*, vol. 31, no. 3, pp. 66-73, 2015.
[http://dx.doi.org/10.1162/DESI_a_00339]

[71]　L. Eadie, and T.K. Ghosh, "Biomimicry in textiles: past, present and potential. An overview", *J. R. Soc. Interface*, vol. 8, no. 59, pp. 761-775, 2011.
[http://dx.doi.org/10.1098/rsif.2010.0487] [PMID: 21325320]

[72]　L. Stevens, M.M. De Vries, M.M. Bos, and H. Kopnina, Biomimicry Design Education Essentials, *International Conference on Engineering Design* vol. vol. 1. University Press: Cambridge, 2019, pp. 459-468.

[73]   O.A. Oguntona, and C.O. Aigbavboa, "Biomimicry principles as evaluation criteria of sustainability in the construction industry", *Energy Procedia,* vol. 142, pp. 2491-2497, 2017.
[http://dx.doi.org/10.1016/j.egypro.2017.12.188]

[74]   R. Rao, "Biomimicry in architecture", *International Journal of Advanced Research in Civil, Structural, Environmental and Infrastructure Engineering and Developing,* vol. 1, pp. 101-107, 2014.

[75]   A. Solga, Z. Cerman, B.F. Striffler, M. Spaeth, and W. Barthlott, "The dream of staying clean: Lotus and biomimetic surfaces", *Bioinspir. Biomim.,* vol. 2, no. 4, pp. S126-S134, 2007.
[http://dx.doi.org/10.1088/1748-3182/2/4/S02] [PMID: 18037722]

[76]   L.I.N. Xuan-yi, "Application of Bionics in Architectural Coatings--Emulsion Paint with Lotus Leaf Effect", *Shanghai Coatings, Z1,* 2005.

[77]   F. Bonviu, "The European economy: from a linear to a circular economy", *Romanian J. Eur. Aff.,* vol. 14, p. 78, 2014.

[78]   G. Michelini, R.N. Moraes, R.N. Cunha, J.M.H. Costa, and A.R. Ometto, "From linear to circular economy: PSS conducting the transition", *Procedia CIRP,* vol. 64, pp. 2-6, 2017.
[http://dx.doi.org/10.1016/j.procir.2017.03.012]

[79]   V. Bach, N. Minkov, and M. Finkbeiner, "Assessing the Ability of the Cradle to Cradle Certified™ Products Program to Reliably Determine the Environmental Performance of Products", *Sustainability (Basel),* vol. 10, no. 5, p. 1562, 2018.
[http://dx.doi.org/10.3390/su10051562]

[80]   S.D. Reay, J.P. McCool, and A. Withell, *Exploring the feasibility of Cradle to Cradle (product) design: perspectives from New Zealand Scientists,* 2011.

[81]   M.E. Peralta Álvarez, F. Aguayo González, J.R. Lama Ruíz, and M.J. Ávila Gutiérrez, "MGE2: A framework for cradle-to-cradle design", *Dyna (Medellin),* vol. 82, no. 191, pp. 137-146, 2015.
[http://dx.doi.org/10.15446/dyna.v82n191.43263]

[82]   M. Braungart, W. McDonough, and A. Bollinger, "Cradle-to-cradle design: creating healthy emissions – a strategy for eco-effective product and system design", *J. Clean. Prod.,* vol. 15, no. 13-14, pp. 1337-1348, 2007.
[http://dx.doi.org/10.1016/j.jclepro.2006.08.003]

[83]   H. Kopnina, "Consumption in environmental education: developing curriculum that addresses cradle to cradle principles", *FactisPax,* vol. 5, pp. 374-388, 2011.

[84]   B. van de Westerlo, J.I. Halman, and E. Durmisevic, "Translate the Cradle to cradle Principles for a Building", In: *International Council for Research and Innovation in Building and Construction. CIB,* 2012.

[85]   H. Kopnina, and J. Blewitt, "Cradle to Cradle (C2C)", *Sustainable Business,* 2nd ed., 2018.
[http://dx.doi.org/10.4324/9781315110172-11]

[86]   A. Cattermole, "How the Circular Economy is Changing Fashion", *AATCC Rev.,* vol. 18, no. 2, pp. 37-42, 2018.
[http://dx.doi.org/10.14504/ar.18.2.2]

[87]   S. van Dijk, M. Tenpierik, and A. van den Dobbelsteen, "Continuing the building's cycles: A literature review and analysis of current systems theories in comparison with the theory of Cradle to Cradle", *Resour. Conserv. Recycling,* vol. 82, pp. 21-34, 2014.
[http://dx.doi.org/10.1016/j.resconrec.2013.10.007]

[88]   https://www.greenroofs.com/projects/ford-motor-companys-river-rouge-truck-plant/

[89]   OECD, *Environment in the Transition to a Market Economy,* 1999.

[90]   S. El Haggar, "Sustainable industrial design and waste management: cradle-to-cradle for sustainable development", *Academic Press,* 2010.

[91]   A.K. Pandey, *Identification and Assessment of Cleaner Production technologies and appropriate technology management strategies and methods in the South African vehicle industry*, 2008.

[92]   J. Coca-Prados, and G. Gutiérrez-Cervelló, *Economic sustainability and environmental protection in Mediterranean countries through clean manufacturing methods.* Springer, 2012.

[93]   J. DuflouandK.Kellens, "Cleaner Production", *The International Academy for Production Engineering,* 2014.

[94]   S.M. El Haggar, "Rural and developing country solutions", In: *Environmental Solutions.*, 2005, pp. 313-400.

[95]   M. Thrane, E.H. Nielsen, and P. Christensen, "Cleaner production in Danish fish processing – experiences, status and possible future strategies", *J. Clean. Prod.,* vol. 17, no. 3, pp. 380-390, 2009.
[http://dx.doi.org/10.1016/j.jclepro.2008.08.006]

[96]   R.J. Cole, *Regenerative design and development: current theory and practice,* 2012.
[http://dx.doi.org/10.1080/09613218.2012.617516]

[97]   P. Mang, and B. Reed, "Designing from place: a regenerative framework and methodology", *Build. Res. Inform.,* vol. 40, no. 1, pp. 23-38, 2012.
[http://dx.doi.org/10.1080/09613218.2012.621341]

[98]   W. Craft, L. Ding, D. Prasad, L. Partridge, and D. Else, "Development of a regenerative design model for building retrofits' , *Procedia Eng.,* vol. 180, pp. 658-668, 2017.
[http://dx.doi.org/10.1016/j.proeng.2017.04.225]

[99]   https://www.archdaily.com/215855/vandusen-botanical-garden-visitor-centre-perkinswil-/513-032_tree_trim

CHAPTER 2

Alternative Building Materials – Road to Sustainability

S. Jeeva Chithambaram[1,*], Pardeep Bishnoi[2], Abhijeet Singh[3], G.S. Rampradheep[4] and C. Ravichandran[5]

[1] *Department of Civil Engineering, Sarala Birla University, Ranchi, Jharkhand, India*

[2] *Clarivate Analytics, Noida, India*

[3] *Department of Mechanical and Automation Engineering, Amity University Jharkhand, Ranchi, Jharkhand, India*

[4] *Department of Civil Engineering, Kongu Engineering College, Perundurai, Tamilnadu, India*

[5] *Department of Environmental Sciences, Bishop Heber College, Tiruchirappalli, Tamilnadu, India*

Abstract: The fundamental and most essential components in building construction are materials. Good design along with the properties of materials (chemical, physical and mechanical) are accountable for a building's material strength. The concept of sustainability in building materials revolves around the development and use of eco-friendly materials that have the same or enhanced properties as compared to that of conventional building materials. However, in the recent past, the use of conventional building materials in building construction has resulted in environmental degradation. Sustainability can be achieved by using industrial waste materials (by-products) and/or recycling and reusing the materials in building construction. Cost (manufacturing and transportation) has been a predominant factor considered while comparing related materials for the same purpose. This chapter discusses the need for alternative sustainable building materials with regard to energy and environmental impact caused by traditional or conventional building materials. Also, this chapter discusses sustainable initiatives carried out by researchers to discover low technology construction techniques. Further, this chapter discusses how alternative building materials can lessen the impact of environmental degradation, resulting in a healthier, cost-effective, and sustainably safe living environment.

Keywords: Alternate building materials, Building Materials, Eco-friendly materials, Sustainability.

---

* **Corresponding author S. Jeeva Chithambaram**: Department of Civil Engineering, Sarala Birla University, Ranchi, Jharkhand, India; E-mail: jeeva.4191@gmail.com

G. Venkatesan, S. Lakshmana Prabu and M. Rengasamy (Eds.)

INTRODUCTION

Housing is considered to be one of the elementary needs of human life and it is a key competent to the sustainable development of a nation. All materials used for construction purposes in buildings are known as building materials. They play a major role in technology as well as in a nation's economy. These materials form the basis of materials found in construction engineering. These materials are used in foundations, floors, walls, beams, roofs, *etc.*

Building materials differ from place to place and the choice of materials depends on certain factors with respect to availability, climate, economy, *etc.* Different materials have been developed around the world with regard to their properties and environmental impact. Also, the fast-growing advancements in construction practices, tools and machinery or technologies may lead to a need for new building materials that could satisfy the increasing demand.

Almost thirty years ago, Sustainable Development gained attention after the 1992 Earth Summit [1] and Brundtland's report entitled "our common Future" [2]. The concept of sustainable development has been defined as development that meets the needs of the present without compromising the ability of future generations to meet their own needs. This concept of sustainability includes enhancing the value of life, thereby letting people live in an atmosphere that is healthy, with better-quality social, economic and ecological conditions. In order to maintain sustainability in recent years, there has been a list of global issues that need to be taken care of. Some of the global issues are listed as follows:

- Weather change,
- Air pollution,
- Lessening of natural resources,
- Biodiversity,
- Waste generation,
- Reduction as well as pollution of water resources, and
- A decline in urban environment.

The emission of carbon dioxide (CO_2) along with other greenhouse gases (GHG) is a result of climate change and global warming. They pose an enormous risk to human wellbeing. In order to contain the hazard, the world needs to reduce the emissions by the use of alternate materials. A huge amount of CO_2 is released into the air throughout the whole life of a building. This comprises the manufacture of building materials, construction of a building, excavation, renovation, possible rehabilitation, and also its ultimate demolition [3].

As per United Nations projections, with regard to the unprecedented rates in a growing population, the total world's population is estimated to reach 9.8 billion in 2050 [4].

CLASSIFICATION OF BUILDING MATERIALS

Generally, building materials are classified based on their chemical composition into three types, namely, inorganic, organic, and composite materials, as given in Fig. (1).

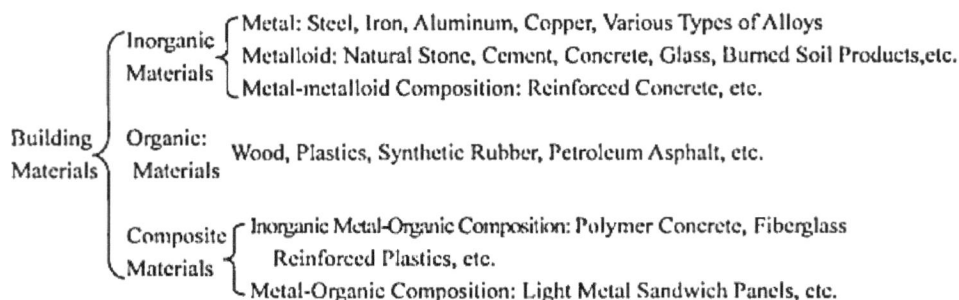

Building Materials

Inorganic Materials
- Metal: Steel, Iron, Aluminum, Copper, Various Types of Alloys
- Metalloid: Natural Stone, Cement, Concrete, Glass, Burned Soil Products,etc.
- Metal-metalloid Composition: Reinforced Concrete, etc.

Organic: Materials
Wood, Plastics, Synthetic Rubber, Petroleum Asphalt, etc.

Composite Materials
- Inorganic Metal-Organic Composition: Polymer Concrete, Fiberglass Reinforced Plastics, etc.
- Metal-Organic Composition: Light Metal Sandwich Panels, etc.

Fig. (1). Classification of building materials.

Also, they are classified as structural materials and functional materials based on the functions of materials. Structural materials are used in beams, columns and plates. Functional materials are used for special purposes like waterproofing, heat-insulating function and ornamental purposes.

Furthermore, building materials can be used based on their physical, chemical and mechanical properties. Some of the physical properties that need to be assessed are density, bulk density, specific weight, specific gravity, porosity, void ratio, water absorption, water permeability, fire resistance, heat conductivity, chemical resistance, *etc.* Some of the mechanical properties of building materials that need to be assessed are strength, hardness, elasticity, plasticity, *etc.*

Some of the traditional building materials used in construction are classified based on their structural, insulation and complementary materials. The most common structural materials are wood, stone, rammed earth, straw bales, clay bricks, cement, *etc.* Some of the most common insulating materials are sheep wool, fibres, *etc.* Plasters, paints and flooring materials are some of the most commonly used complementary materials.

Stone is one of the oldest construction materials, which is considered to be highly durable along with a low maintenance cost. After stone, wood is considered to be the oldest material used by humans for construction and building purposes [5]. Nowadays, cement concrete is considered to be the main component of building construction. Cement concrete is considered to be a homogeneous mixture comprising cement, fine aggregate, coarse aggregate and water in a traditional form. All these construction materials form the basis of a building and are responsible for the life-cycle of a building.

Need For Alternative Building Materials

Day to day need for housing has been increasing at a rapid pace both in rural as well as urban areas. This increase in demand for building materials for construction will lead to the depletion of natural resources, high energy consumption, CO_2 and greenhouse gas emissions, *etc.* (Fig. **2**) displays the data gathered by the International Energy Agency in 2020, which evaluates the carbon dioxide (CO_2) emissions from the burning of coal, natural gas, oil, and other fuels, including industrial waste and non-renewable municipal waste. Currently, India ranks the third and contributes to 7% of the total world's CO_2 emissions, *i.e.*, 2.2GT.

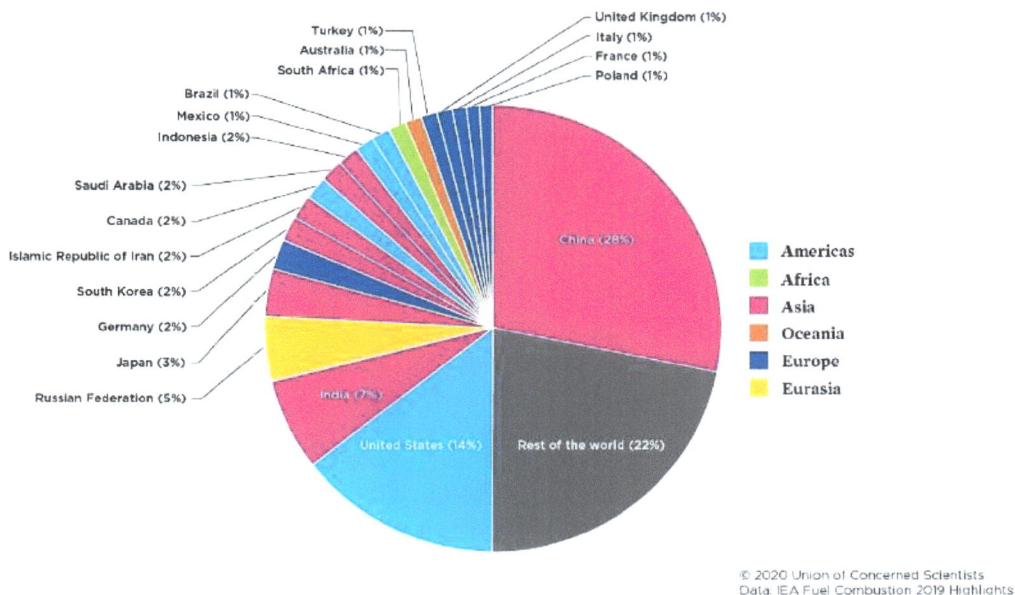

© 2020 Union of Concerned Scientists
Data: IEA Fuel Combustion 2019 Highlights

Fig. (2). Each country's share of CO_2 emissions [6].

Due to the continuous depletion of resources in future, there will be a change in construction towards building materials that are easily and locally available as well as that consume less energy [7]. Easy availability in near-by places may reduce the transportation cost of the material. Yet in a few cases, the cost of transportation may be compensated if materials with a comparatively better performance is available elsewhere.

Alternative Building Materials

In the present-day scenario, cement concrete is the fundamental material used for construction in developing nations. Usually, cement concrete comprises cement, fine aggregate, coarse aggregate and water. Recently, admixtures (both mineral as well as chemical) are added to concrete to alter its binding mechanism. Due to its increasing demand, researchers are in search of alternates that may contribute towards sustainability without hindering the overall performance. These alternate building materials should be sound enough in terms of function, environmental benefits, cost, affordability, accessibility, handling aspects and also should be non-hazardous. In addition, these materials should follow the guidelines of a country's regulations, with a clear bind ability to the other general ingredients. Some of the notable materials in research are supplementary materials to cement, alternative aggregates, new kinds of admixtures either to smoothen the traditional function or to catalyse the reaction in a better way.

Materials which possess the functions of conventional cementitious materials are classified as Supplementary Cementitious Materials (SCM's). Some of the examples of most commonly used SCMs are Fly Ash, Silica Fume, Ground Granulated Blast Furnace Slag (GGBS), Metakaolin, Quarry Dust, *etc.*, These materials can be used as alternatives when added to Ordinary Portland Cement. ASTM Classifications of Class C and Class F as an alternative to the fine powder material function better in respect to the field of applications. Slags are gaining importance in the recent past due to the presence of CaO. Table **1** shows the typical chemical and physical properties of Cement and Supplementary Cementitious Materials.

Many of the alternative building materials like the bio-based and earthen building materials turn out to be biodegradable and do not yield hazardous by-products [16 - 27]. Building materials having renewable organic plant and animal constituents can be categorized under bio-based building materials. They may be of agricultural origin, animal waste or biological wastes [16]. Some of the most common examples of bio-based materials are bamboo, straw bales, fibre crops, agricultural residues and plant seed oils.

Table 1. Typical chemical and physical properties of cement and supplementary cementitious materials [8 - 15].

Physical Properties	Cement (OPC)	Silica Fume	Fly Ash	Quartz Powder	Glass Powder	GGBS	Rice Husk Ash
Median Particle Size (μm)	10-45	0.1-1	13-40	5-45	0.1-45	6.5-45	3.8-10
BET Surface Area (cm²/g)	3,300-3,800	140,000-200,000	12,000-97,000	7,500	4,000-4,630	3,720-8,000	2,500,000-3,040,000
Density (kg/m³)	3,150-3,160	2,220-2,260	2,250-2,560	2,700	2,490-2,579	1,200-2,780	90-490
Chemical Compound (%)	-	-	-	-	-	-	-
SiO_2	20.07-22	92.85-95	53.09-53.5	99.24-99.4	63.79-72.5	28.3-35.34	87.32-94.95
Al_2O_3	4.47-6.6	0.61-0.9	20.04-24.80	0.05-0.35	0.4-2.62	11.59-13.6	0.22-0.39
Fe_2O_3	2.8-2.91	0.60-0.94	8.01-8.66	0.017-0.04	0.2-1.42	0.35-0.62	0.26-0.67
CaO	60.1-63.89	0.30-0.39	2.44-3.38	0.03-0.28	9.7-12.45	38.4-41.99	0.48-0.67
MgO	3.03-3.3	0.9-1.58	1.94-2.25	0.01	2.73-3.3	7.2-8.04	0.28-0.44
SO_3	2.91	-	0.23-0.6	-	-	0.23-7.4	-
Alkali ($Na_2O+0.658 K_2O$)	0.68	1.07	3.27	-	12.5-14.33	-	1.97-3.08612
Loss on Ignition	1.0-2.6	2.1-3	1.2-3.59	0.06	0.36-0.45	-	0.85-2.10

Bamboo is a sustainable alternative to the conventional building material with excellent mechanical properties, light in weight, flexible in nature and relatively less cost. Bamboo can be used in interiors of the house to laminate flooring, chip boards, *etc.* [16].

Bamboo, straw, hemp, rye stalk, linseed, sunflower stalk, rice husk, and cork are some of the rapid renewable resources that are utilized in Alternative Building Materials [17]. Sheep wool is an example of animal based renewable material used for insulation [18].

On the other hand, building materials used for earthen building like hydrated lime, clay, mud bricks, compressed earth blocks and rammed earth have been used for several years all over the world [28, 29]. Presently, there is a increasing attention in finding out sustainable replacements to traditionally used concrete, brick and wood. Numerous new publications have presented queries regarding the soil characterization, manufacturing process and testing of materials [30 - 35].

In recent days, geopolymers or alkali activated materials or geocements [36 - 38] are gaining importance among researchers. Alkali-activated Materials are materials that come under the concept of a reactive solid material that toughens under an alkaline activator solution. AAMs may be well-defined as three dimensionally interacted amorphous to semi-crystalline aluminosilicate materials. Slags and Fly ash are the most chosen and consistent precursor materials for alkali-activation. The knowledge of alkali-activated materials as an alternative to Portland cement has been documented since at least 1908. Purdon was the foremost researcher who laid the scientific foundation for AAMs in 1940 in Belgium [39]. Pluhowsky [40] (1959) was the first and the most well-known researcher who focused on alkali-carbonate activation of metallurgical slags.Some of the most extensively and usually used precursors materials to produce alkali activated materials are:

- Blast Furnace Slag
- Fly Ash

Also, Red mud, Rice Husk Ash, Metakaolin, Palm Oil Fuel Ash, Ceramic Waste, Marble powder, Waste Glass, *etc.*, are some of the other precursors that have been utilized to develop alkali activated materials, most of which have been unexplored. A combination of Sodium Hydroxide/Potassium Hydroxide and Sodium Silicate/Potassium Silicate is the most usually and most extensively utilized alkaline activator liquid. Among these, sodium hydroxide is the cheapest and most available alkaline hydroxide. Therefore, it has been the most usually utilized activator. These are an alternative to the normal cement and possess superior properties such as fineness, thermal resistivity, resistance to expansion and strength. The usage of such alternates not only improves the physical properties but also plays an appreciable role in conserving the environmental conditions especially in controlling the emission of CO_2 into the atmosphere, as the carbon footprint is the main problem during the production of normal type cement. In geopolymers, since fly ash or slag is used as the source material, considerable amount of CO_2 emission is reduced. Also, usage of a waste material as the primary source material reduces the energy consumption by a considerable amount.

Similarly, Alccofine 1203 performs superior to the SCMs in particular to the action of silica fume-based materials in the concrete. High performance-based concrete could be achieved by satisfying the significant durability test especially the permeability, and thus improves the pore filling effect, dense matrix as because of the perfect hydration with respect to the time. Thus, the interfacial transition zone will be maintained in order for a long run compared to the other kinds of cement.

A special type of material named Glazed Iso Ball normally coined as GIB is a light weight material that can be used as a replacement to aggregates. The nature of GIB will be of spherical ball like bubbles and the closed cell arrangement of the material will possess low water absorption, high mechanical strength and appreciable fire resistance property (Rampradheep G.S. *et al.*, IPR Design Journal 2020).

The recycled aggregates in processed and graded form can be the other substitute for the fine and coarse aggregates depending upon the particle size characteristics. Construction & Demolition Debris normally to be called as C & D debris can be used as an alternative to conventional aggregates in concrete, paves the solution for waste management. CO_2 sequestration in the C & D based concrete or mortar is an advanced technology which helps to form Calcium Carbonate and hence a dense microstructure could be ensured [41].

Marble powder is the other alternate material that is obtained from the waste marble and is used in concrete either as part replacement to cement or fine aggregates. Industrial waste materials such as ceramic dust, quarry dust, siliceous powder, slag particles *etc.*, in a synthesized form can be used as a part replacement to conventional fine aggregates. The materials used will possess the filling ability and shear resistance results in good binding property with the surrounding additives for concrete.

Papercrete is a type of lightweight concrete consisting of Portland cement along with re-pulped paper. It is considered to be an environmentally friendly material with lower thermal conductivity than concrete. Hempcrete is a bio-composite building material consisting of hemp hurds along with lime binder [42].

Some of the best examples, as shown in Fig. (**3**) indicating the use of SCM's as partial replacement of cement in green buildings include the Bank of America Tower, New York, USA (45% blast furnace slag); The Helena, New York, USA (45% blast furnace slag); and the San Francisco Federal Building, California, USA (50% blast furnace slag) [43].

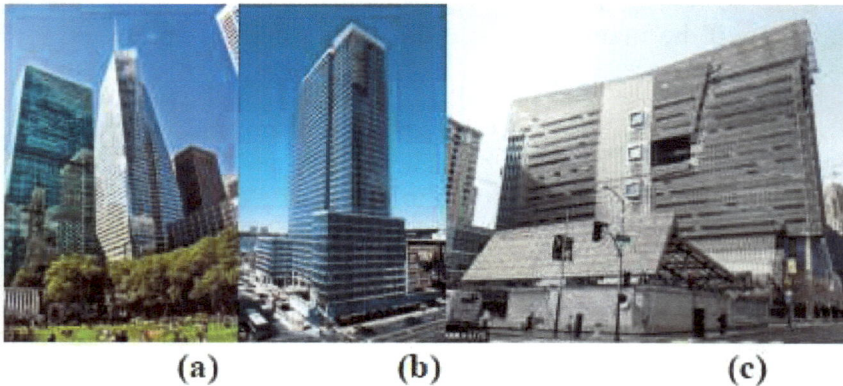

(a)   **(b)**   **(c)**

Fig. (3). Buildings using SCMs as partial replacement of cement (**a**) The Bank of America Tower, (**b**) The Helena, (**c**) San Francisco Federal Building [43].

The EcoARK in Taipei is a type of building featuring recycled plastic material as the building skin [44]. Polli-Bricks are used in this structure to cover he façades and walls of the building. "Polli-Bricks" are a hollow, translucent building block made of recycled polyethylene terephthalate (PET).

Another best example of the recycle and reuse principle is the mega restoration of Jackie Chan Stuntman Training Centre as shown in Fig. (4). For restoration of the half built, existing structures left to decay including a movie complex, shopping mall and stadium sized event space, many upcycled materials like DVD's, tyres and polyurethane plastic bags were used to restore the outer- façades, artificial turf and flooring, respectively [45].

Fig. (4). Jackie Chan Stuntman Training Centre [45].

Concept of Sustainability in Building Materials

Currently, the concept of sustainability has become an interesting field of research among researchers. The main focus is to tackle the global warming indices along with the climate change associated. Various nations across the globe have adopted principles and policies to control the increasing effects of global warming and to improve the present scenario. This applies for the construction industry as well. Some of the strategies to adopt sustainability in the construction industry includes:

- Reduce energy consumption,
- Balanced use of raw materials or natural resources,
- Regular monitoring and controlling the emissions.

Systematic approach towards the above-mentioned measures must be applied while selection of materials for building and construction activities. Some of the major problems faced while selection of building materials is sourcing of materials, their performance and their cost. The concept of sustainability in building materials can be summarized as follows:

- Use of renewable energy to manufacture, produce and transport materials.
- Reduce the use of raw materials or natural resources.
- Use of locally available materials thereby reducing the transportation cost.
- Use of economy effective construction skills and techniques.
- Recycle of possible building materials.
- Reuse of possible building materials or use of recycled building materials.
- Increase the overall performance and service life of a building.
- Use of composite building materials.

CONCLUSION

Sustainability in construction industry is much needed for the betterment of the society. The fundamental concept of sustainability paves way for alternative to building materials. Some of the benefits of using alternative building materials as compared to that of the conventionally used building materials are:

- Reduced greenhouse gas emissions
- Improved strength
- Improved durability
- Improved total service life
- Improved functionality
- Improved environmental performance

- Reduced cost
- Reduced toxicity
- Reduced air pollution
- Improved living environment
- Improved recycling potential, *etc.*

CONSENT FOR PUBLICATION

Not applicable.

CONFLICT OF INTEREST

The author declares no conflict of interest, financial or otherwise.

ACKNOWLEDGEMENT

Declared none.

REFERENCES

[1]   *Report of the World Commission on Environment and Development: Our Common Future.* http://www.un-documents.net/wced-ocf.htm

[2]   O. Ortiz, F. Castells, and G. Sonnemann, "Sustainability in the construction industry: A review of recent developments based on LCA", *Constr. Build. Mater.,* vol. 23, no. 1, pp. 28-39, 2009. [http://dx.doi.org/10.1016/j.conbuildmat.2007.11.012]

[3]   M.J. González, and J. García Navarro, "Assessment of the decrease of CO2 emissions in the construction field through the selection of materials: Practical case study of three houses of low environmental impact", *Build. Environ.,* vol. 41, no. 7, pp. 902-909, 2006. [http://dx.doi.org/10.1016/j.buildenv.2005.04.006]

[4]   UN, *2017 Revision of World Population Prospects. ST/ESA/SER.A/399. Department of Economic and Social Affairs Population Division.* United Nations: New York, 2017.

[5]   M. Kozlovska, Z. Strukova, and P. Kaleja, "Methodology of cost parameter estimation for modern methods of contruction based on wood", *Proc. Engi,* vol. 108, pp. 387-393, 2015.

[6]   https://www.ucsusa.org/resources/each-countrys-share-co2-emissions

[7]   J.C. Morel, A. Mesbah, M. Oggero, and P. Walker, "Building houses with local materials: means to drastically reduce the environmental impact of construction", *Build. Environ.,* vol. 36, no. 10, pp. 1119-1126, 2001. [http://dx.doi.org/10.1016/S0360-1323(00)00054-8]

[8]   L.M. Federico, and S.E. Chidiac, "Waste glass as a supplementary cementitious material in concrete – Critical review of treatment methods", *Cement Concr. Compos.,* vol. 31, no. 8, pp. 606-610, 2009. [http://dx.doi.org/10.1016/j.cemconcomp.2009.02.001]

[9]   K. Ganesan, K. Rajagopal, and K. Thangavel, "Rice husk ash blended cement: Assessment of optimal level of replacement for strength and permeability properties of concrete", *Constr. Build. Mater.,* vol. 22, no. 8, pp. 1675-1683, 2008. [http://dx.doi.org/10.1016/j.conbuildmat.2007.06.011]

[10] K. Habel, and P. Gauvreau, "Response of ultra-high performance fiber reinforced concrete (UHPFRC) to impact and static loading", *Cement Concr. Compos.,* vol. 30, no. 10, pp. 938-946, 2008.

[http://dx.doi.org/10.1016/j.cemconcomp.2008.09.001]

[11]   G. Long, X. Wang, and Y. Xie, "Very-high-performance concrete with ultrafine powders", *Cement Concr. Res.*, vol. 32, no. 4, pp. 601-605, 2002.
[http://dx.doi.org/10.1016/S0008-8846(01)00732-3]

[12]   A.E. Naaman, and K. Wille, "The path to ultra-high performancefiber reinforced concrete (UHPFRC)", 3rd *Int. Symp.* on UHPC, M. Schmidt, E. Fehling, C. Glotzbach, S. Fröhlich, and S. Piotrowski, eds., Kassel University Press, Kassel, Germany, 3–15, 2012.

[13]   V.G. Papadakis, E.J. Pedersen, and H. Lindgreen, "An AFM-SEM investigation of the effect of silica fume and fly ash on cementpastemicrostructure", *J. Mater. Sci.*, vol. 34, no. 4, pp. 683-690, 1999.
[http://dx.doi.org/10.1023/A:1004500324744]

[14]   N. Schwarz, H. Cam, and N. Neithalath, "Influence of a fine glass powder on the durability characteristics of concrete and its comparison to fly ash", *Cement Concr. Compos.*, vol. 30, no. 6, pp. 486-496, 2008.
[http://dx.doi.org/10.1016/j.cemconcomp.2008.02.001]

[15]   J. Willey, *"Use of ultra-high performance concrete to mitigate im-pact and explosive threats"*, Missouri Univ. of Scienceand Technology, Rolla, MO., Master's thesis, 2013.

[16]   M. Calkins, *Materials for Sustainable Sites: A Complete Guide to the Evaluation, Selection, and Use of Sustainable Construction Materials.* John Wiley & Sons: Hoboken, NJ, USA, 2009.

[17]   E Latif, M Lawrence, A Shea, and P Walker, *"In situ* assessment of the fabric and energy performance of five conventional and non-conventional wall systems using comparative coheating tests", *Build Environ*, vol. 109, no. 6, pp. 68-81, 2016.

[18]   J. Zach, A. Korjenic, V. Petránek, J. Hroudová, and T. Bednar, "Performance evaluation and research of alternative thermal insulations based on sheep wool", *Energy Build.*, vol. 49, pp. 246-253, 2012.
[http://dx.doi.org/10.1016/j.enbuild.2012.02.014]

[19]   S. Halliday, *Sustainable Construction.* 1st ed. Butterworth-Heinemann: Oxford, UK, 2008.
[http://dx.doi.org/10.4324/9780080557168]

[20]   P. van der Lugt, A.A.J.F. van den Dobbelsteen, and J.J.A. Janssen, "An environmental, economic and practical assessment of bamboo as a building material for supporting structures", *Constr. Build. Mater.*, vol. 20, no. 9, pp. 648-656, 2006.
[http://dx.doi.org/10.1016/j.conbuildmat.2005.02.023]

[21]   A. Utama, and S.H. Gheewala, "Influence of material selection on energy demand in residential houses", *Mater. Des.*, vol. 30, no. 6, pp. 2173-2180, 2009.
[http://dx.doi.org/10.1016/j.matdes.2008.08.046]

[22]   K.D. Flander, and R. Rovers, "One laminated bamboo-frame house per hectare per year", *Constr. Build. Mater.*, vol. 23, no. 1, pp. 210-218, 2009.
[http://dx.doi.org/10.1016/j.conbuildmat.2008.01.004]

[23]   C.P. Hoang, K.A. Kinney, and R.L. Corsi, "Ozone removal by green building materials", *Build. Environ.*, vol. 44, no. 8, pp. 1627-1633, 2009.
[http://dx.doi.org/10.1016/j.buildenv.2008.10.007]

[24]   L. Jayanetti, and P. Follet, "Building with sustainable forest products", *Struct. Eng.*, vol. 81, pp. 14-17, 2003.

[25]   S. Paudel, and M. Lobovikov, "Bamboo housing: market potential for low-income groups", *J. Bamboo Rattan,* vol. 2, no. 4, pp. 381-396, 2003.
[http://dx.doi.org/10.1163/156915903322700412]

[26]   B. Isik, and T. Tulbentci, "Sustainable housing in island conditions using Alker-gypsum-stabilized earth: A case study from northern Cyprus", *Build. Environ.*, vol. 43, no. 9, pp. 1426-1432, 2008.
[http://dx.doi.org/10.1016/j.buildenv.2007.06.002]

[27]   C.H. Kouakou, and J.C. Morel, "Strength and elasto-plastic properties of non-industrial building materials manufactured with clay as a natural binder", *Appl. Clay Sci.,* vol. 44, no. 1-2, pp. 27-34, 2009.
[http://dx.doi.org/10.1016/j.clay.2008.12.019]

[28]   F. Collet, L. Serres, J. Miriel, and M. Bart, "Study of thermal behaviour of clay wall facing south", *Build. Environ.,* vol. 41, no. 3, pp. 307-315, 2006.
[http://dx.doi.org/10.1016/j.buildenv.2005.01.024]

[29]   J. Kennedy, Ed., *Building without Borders: Sustainable Construction for the Global Village.* New Society Publishers: Gabriola Island, Canada, 2004.

[30]   M. Hall, and D. Allinson, "Assessing the moisture-content-dependent parameters of stabilised earth materials using the cyclic-response admittance method", *Energy Build.,* vol. 40, no. 11, pp. 2044-2051, 2008.
[http://dx.doi.org/10.1016/j.enbuild.2008.05.009]

[31]   C. Jayasinghe, and N. Kamaladasa, "Compressive strength characteristics of cement stabilized rammed earth walls", *Constr. Build. Mater.,* vol. 21, no. 11, pp. 1971-1976, 2007.
[http://dx.doi.org/10.1016/j.conbuildmat.2006.05.049]

[32]   Q.B. Bui, J.C. Morel, B.V. Venkatarama Reddy, and W. Ghayad, "Durability of rammed earth walls exposed for 20 years to natural weathering", *Build. Environ.,* vol. 44, no. 5, pp. 912-919, 2009.
[http://dx.doi.org/10.1016/j.buildenv.2008.07.001]

[33]   J.C. Morel, A. Pkla, and P. Walker, "Compressive strength testing of compressed earth blocks", *Constr. Build. Mater.,* vol. 21, no. 2, pp. 303-309, 2007.
[http://dx.doi.org/10.1016/j.conbuildmat.2005.08.021]

[34]   B.V. Venkatarama Reddy, and A. Gupta, "Influence of sand grading on the characteristics of mortars and soil–cement block masonry", *Constr. Build. Mater.,* vol. 22, no. 8, pp. 1614-1623, 2008.
[http://dx.doi.org/10.1016/j.conbuildmat.2007.06.014]

[35]   V. Maniatidis, and P. Walker, "Structural capacity of rammed earth in compression", *J. Mater. Civ. Eng.,* vol. 20, no. 3, pp. 230-238, 2008.
[http://dx.doi.org/10.1061/(ASCE)0899-1561(2008)20:3(230)]

[36]   J. Davidovits, "Proceedings of the 1st European conference on soft mineralurgy "Geopolymer '88", *Saint Quentin.,* vol. 12, pp. 25-48, 1988.

[37]   P. Krivenko, *Proceedings of the tenth international congress on the chemistry of cement, Go¨theborg.,* pp. 4iv046-4iv050, 1997.

[38]   P. Krivenko, "Alkaline cements and concretes", Proceedings of the 2nd international conference on alkaline cements and concretes. Oranta Ltd, Kyiv 1999.

[39]   A.O. Purdon, "The action of alkalis on blast furnace slag", *J. Soc. Chem. Ind.,* vol. 59, pp. 191-202, 1940.

[40]   V.D. Glukhovsky, *Soil Silicates.* Gostroiizdat Publish: Kiev, USSR, 1959.

[41]   G.S. Rampradheep, S. Anandakumar, and M. Diwakar, *Construction Blocks from C & D Debris Using the Innovative CO_2 Sequestration Technique. Sustainable Construction and Building Materials. Lecture Notes in Civil Engineering.* vol. Vol. 25. Springer: Singapore, 2019.

[42]   A. Sutton, D. Black, and P. Walker, *Hemp lime: an introduction to low-impact building materials. Building Research Establishment: Information Paper IP 14/11.* IHS BRE Press: UK, 2011.

[43]   J. Gonchar, "Building even better concrete", *Archit. Rec.,* vol. 195, no. 2, pp. 143-149, 2007.

[44]   *Miniwiz EcoArk.,* 2019. http://www.miniwiz.com/solution_detail.php

[45]   *Miniwiz Jackie Chan Stuntman Training Center,* 2019. http://www.miniwiz.com/solution_detail.php?id=39

CHAPTER 3

Smart Waste Management to Enrich Cleanliness and Reduce Pollution in the Environment

N. Shenbagavadivu[1,*], **M. Bhuvaneswari**[2] and **G. Jenilasree**[3]

[1] *Department of Computer Applications, University College of Engineering, BIT Campus, Anna University, Tiruchirappalli, India*

[2] *Department of ECE, University College of Engineering, BIT Campus, Anna University, Tiruchirappalli, India*

[3] *Reaserach Scholar, University College of Engineering, BIT Campus, Anna University, Tiruchirappalli, India*

Abstract: The forecast of waste generation is an essential step for adequate waste management planning since it involves various factors that can affect waste trends. Due to over-population in urban areas, the rate of garbage production has been increasing rapidly. To simplify the process, a proposed solution for a smart solid waste management system has been implemented. The proposed smart bin has a faster and more intelligent separation process of the waste material.

Keywords: Ardunio Uno Board, Compressor, Infra-red sensor, LED, Solar battery backup, Solar panel, Temperature sensor, Ultra sonic sensor.

INTRODUCTION

Various critical factors, including its prime location, clean ambience, abundant water, polite people, religious harmony, good rail, road, and air connectivity, have all contributed to a modest increase in the number of tourists visiting the city. In the coming decades, solid waste management (SWM) will be an important parameter in accurately judging a civic body's accomplishment. Solid waste management is categorized into various forms, namely bio-degradable waste, non-bio-degradable waste, and hazardous waste [1]. In recent days, COVID-19, also known as the coronavirus, has been said to lead to the life thread of human life since the infectious diseases are spread to humans *via* sneezing, coughing, touching, saliva, *etc*. In the hospital, wastes such as blood bags, syringes, cotton, dresses, masks, and so on are not properly collected or dumped.

---

[*] **Corresponding author N. Shenbagavadivu:** Department of Computer Applications University College of Engineering, BIT Campus Anna University Trichy, India; E-mail: kshenthu@gmail.com

G. Venkatesan, S. Lakshmana Prabu and M. Rengasamy (Eds.)
All rights reserved-© 2022 Bentham Science Publishers

Due to this improper collection of waste materials, the virus can be spread to the environment, and the number of affected patients will increase rapidly. All three types of waste products are collected through the garbage bin at a common place in an area/street [2]. The accumulation or overflow of garbage from the bin leads to the environment getting polluted due to the lack of collection of garbage from the streets. This typically happens because the local collector does not know about the garbage naturally produced or garbage piled in the local streets. In the current process, a responsible person has to wander through the different spots and carefully check the places for waste collection. This is a somewhat complex and time-consuming process [3]. In the present day, a waste management system is not as efficient as it should have been, taking into consideration the advancements in technology that have naturally arisen in recent years. There is no surety about the management/clearing of waste at all the places. There is a pressing need to maintain the city's hygiene and protect the environment from degradation to avoid unhealthy environments. To prevent the accumulation or overflow of garbage waste in the bin, a new approach to smart solid waste management (SSWM) systems is proposed [1]. It is a necessary step forward for making the waste collection process automatic and efficient in nature [4, 5].

CURRENT STATUS OF THE GARBAGE COLLECTION

Solid Waste Management in Italy

About 58.5 million people live in Italy, a European country. It has three macro-geographical areas, namely: North and South. These three regions consist of nearly 20 regions. In 2019, Italy generates about 51.1 million metric tonnes of municipal solid waste (MSW). In total MSW production, 32.7% is the source-separated collection of recyclables and compostable [6]. The North and South each reached a value of 55.5%, 28.3%, and 18.1%, respectively. MSW combusted at waste-to-energy (WTE) ranged from about 1.6 million tonnes to 3.5 million tons. The North sends to WTE facilities the largest quantity of MSW and RDF (refuse derived fuel) [2]. About 9 million tonnes of MSW would be managed by the mechanical-biological treatment in 2019, increasing the production of compost and bio-stabilized, dry fraction, or RDF, by 40% and 80%, respectively. The number of landfills in Italy decreased from 657 to 401 between 2011 and 2019. In the south of Italy, most of the landfills are present in areas in which the landfills are not uniform [7]. From 2000 to 2004, the use of landfills decreased from 72.4% to 51.9%, but Italy remains the principal method of disposal. WTE increased from 8.5% to 9.7% [3]. Between 2011 and 2019, the use of mechanical-biological treatment and composting remained constant at about 28%.

Solid Waste Management in Switzerland

In Switzerland, municipal solid waste (MSW) has been increasing from year to year. It reached 720 kg per person in 2007. Swiss recycling rates are among the highest in the world. Their recycling principle is applied very swiftly everywhere. Electricity and heat are generated by incinerating wastes in the incinerating process, which accounts for about 2% of the country's final energy requirements [8]. The level of environmental pressure has been reduced by the Swiss Confederation's waste management policy. The introduction of high waste management standards leads to the creation of a highly effective infrastructure. Waste producers are responsible for the cost of disposal, which is made by the financing system. Highly effective waste policies are even insufficient in reducing the country's overall consumption of resources [4]. Hazardous waste accounts for approximately 6% of total waste. For special reprocessing, 1.2 million tonnes of hazardous waste are consigned, or wastes are exported in line with the control of Tran's boundary movements of hazardous wastes and their disposal [9]. Over 1 billion francs will be spent on remediating disused hazardous waste landfill sites.

Indian Scenario and Technological Gap

Garbage is waste generally thrown out of our homes, offices, shops and restaurants. In our country, almost half of it consists of rotting vegetables and food matter [3]. Besides, it also contains paper, plastic, glass, rubber, leather, coal, metal, rags, toxic materials (such as batteries, pesticides, paints, and chemicals), building materials, and soil. According to the Central Pollution Control Board (CPCB), the average Indian generates about 490 tons of waste per day [10]. Although the per capita waste is low compared to western countries, the volume is huge. The generation of solid waste in Indian cities has been estimated to grow by 1.3 percent annually. The expected generation of waste in 2025 will therefore be around 700 tons per capita per day [5]. Considering that the urban population of India is expected to grow to 45 percent from the prevailing 28 percent, the magnitude of the problem is likely to grow even larger unless immediate steps are taken [11]. While the quantity of solid waste generated by society is increasing, the composition of waste is becoming more and more diversified, with the increasing use of packaging materials made of both paper and plastic. Thirty years ago, the composition of solid waste generated by the Indian farmer was characterized by one-fifth non-biodegradable waste and four-fifths biodegradable waste [6]. This ratio is about to reverse; currently, only 40% of all solid waste is biodegradable, while 60% is non-biodegradable.

Technological Gap - Operational/Service Level

- Area with limited service coverage
- Low waste collection efficiency
- Low waste recovery/processing
- Low scientific disposal of waste

Capacity and Capability Issues - Technological Gap

- Lack of adequate manpower
- Lack of technical expertise
- Lack of awareness mechanisms or community participation

THE CRITICAL IMPORTANCE OF THE PROPOSED SMART WASTE MANAGEMENT SYSTEM

The main aim of this proposed method is to keep the region clean to attract more tourists. Garbage collection vehicles and garbage bins play a vital role in this cleanliness campaign. If the garbage from households is not collected at regular intervals, the waste bins will overflow, leading to an unhygienic environment, which makes it unpleasant not only for the tourists but also for the pilgrims and local residents [12].

The detection, monitoring, and management of waste are one of the primary problems of the present era. The traditional way of manually monitoring the waste in waste bins is a complex, cumbersome process that utilises more human effort, time, and cost, which is not compatible with present day technologies in any way. It has become mandatory to find a scientific way of managing garbage. The overall work flow of the proposed model is shown in Fig. (1). It is high time that we implement some smart method to remove solid hazardous waste from households at regular intervals. Hence, the aim of this project is to utilise science and technology to remove solid waste from households to make the city smart and hygienic [8].

To collect the household garbage on time, GPS technology and mobile applications are used. For the automation of garbage collection systems, smart bins play a vital role. The smart bin is designed with the help of GPRS (GPS), an Ardunio UNO microcontroller, sensors, and a solar power panel along with the compaction unit. Every time garbage is collected, information in smart bins is monitored and necessary action is taken. In this case, the automation of vehicle tracking and allocation is used, which will reduce time and money. This proposed system also has the provision for setting up a GPS module in every household so that the arrival of the garbage van is intimated to its inmates, which eases the waste collection process [13]. The project also aims to develop a simple android mobile application on both the client and server side for daily and constrained garbage collection processes. The mobile application shall be developed in such a way that the notification with an alert sound about the arrival of the garbage van is made even if the mobile is in silent mode. Furthermore, signal boosters will be installed in garbage trucks to boost mobile signal in areas with poor network coverage. This project also proposes to set up a mobile set-top box that will be designed and set up in households to notify them with a beep sound about the arrival of garbage trucks. This set-top box serves multiple purposes. The use of sensors like ultrasonic and thermal sensors sends information about the level and fire alarm to the centralised smart bin monitoring system, which will be made available on the server side. The database on garbage vans, employees, *etc*. will be stored in a cloud environment. All the above technologies are interconnected by means of Iota. Thus, this proposed project intends to produce a clean as well as green atmosphere and prevent household waste from being hazardous under the Clean India Mission.

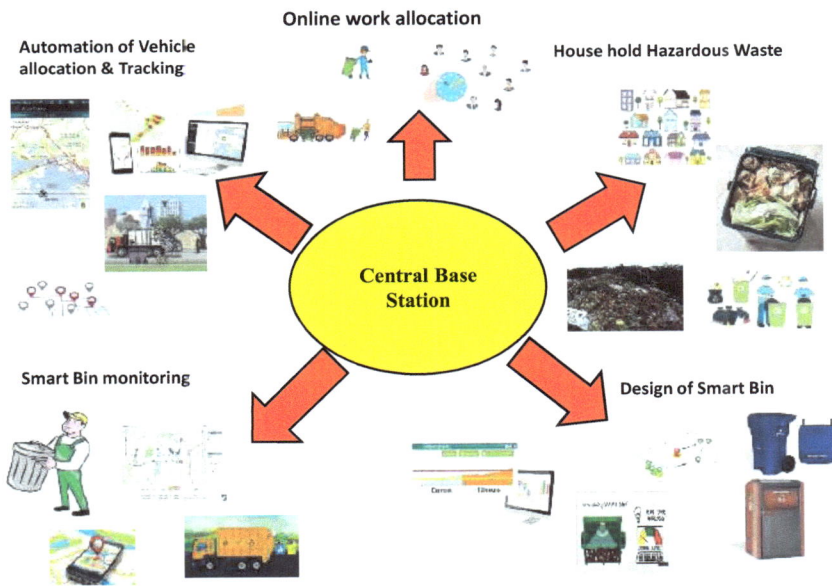

Fig. (1). Architecture of SSWM.

THE PROPOSED MODEL'S TECHNOLOGIES

Technologies like GPS, IOT, Ardunio UNO microcontroller programming kit, sensors, solar power panel, web services, cloud computing, and signal booster are to be used.

- Our proposed system comprises the installation of trucks and bins with sensors like ultra-sonic, optical sensors and thermal sensors, which can constantly record the amount of waste collected at various locations, detect fire if caused accidently in the bins, and create a database.
- This database can be instrumental in understanding, analysing, and predicting waste production patterns and eventually managing waste more efficiently.
- Our proposed smart bins shall avoid the overflow of garbage to create a clean environment through the use of sensors that eventually sense the level of garbage and send messages to the control room [14].
- Garbage trucks in the surrounding area can be located using a GPS device installed in them and instructed to collect trash from that specific bin, as used in cab booking services.
- The filling and laundry time of smart bins will also be wide-ranging, thus making empty and fresh dustbins accessible to people.
- The compaction mechanism should be employed in smart bins to compress the waste put inside them. This can minimise the number of bins used. In our proposed work, one smart bin is equivalent to eight ordinary bins that we currently use.
- Non-biodegradable and biodegradable wastes must be collected in separate bins for further processing. An automated voice system will be used to reject non-biodegradable waste such as plastics if they are placed in the wrong bin.
- To find the shortest route, prediction and route algorithms are used. It will reduce the workforce, money, the number of trucks required for cleaning and the amount of fuel. It also aims at creating a clean as well as green environment, as it will reduce fuel consumption and, in turn, reduce air pollution.

The mobile set-top box further simplifies the garbage collection mechanism by giving alert sounds on the arrival of garbage trucks, ensuring no garbage is left out. Even if the mobile is in silent mode, alert sounds can be produced.

- The signal booster devices will be installed in the garbage trucks to boost the mobile signal level in the regions with low network coverage.
- The use of solar power panels for smart bins is an initiative to green India.Our project is a small step towards such green technology, which can be further extended to the city and so on.

PROPOSED MODEL

The main objective is to integrate information technology with the existing waste management system, which has a huge impact on the condition of the waste sector in India. The basic objective is to minimise time consumption and investment, maximise the quality of the service provided, and ensure no waste is left out in households as hazardous. For instance, the collection system can be improvised if the quantity and quality of the waste collected are regularly monitored. This knowledge will eventually help in better planning of collection routes, types of vehicles to be used, and identification of critical areas that probably need special attention. The long-term objective is, therefore, to reduce the environmental degradation caused by solid waste.

Particular Goals

- Smarter working for effective waste management through smart bins.
- Cost-effective Smart Solid waste by compaction of waste in smart bins. Hence, one smart bin is equivalent to eight ordinary bins.
- Cost-effective in terms of reducing the fuel costs by implementing the shortest route algorithm.
- Time consumption reduction by means of automation of work allotment and online monitoring of Smart Bins.
- Future predictive analysis of waste management through maintaining a centralised database.
- Sensors to avoid bin overflow.
- Development of mobile applications and installation of mobile set-top box devices for efficient garbage removal from households.
- It aims to create a clean and green environment with no trash left exposed to danger.
- Reduction in fuel consumption through vehicle monitoring, which in turn reduces air pollution.

METHODOLOGY AND SCIENTIFIC FOUNDATION

The proposed system combines traditional garbage collection systems with advanced information technology, including smart sensors, the Ardunio UNO microcontroller, a GPS device for vehicle tracking, a mobile application, and a mobile set-top box to provide notification of garbage truck arrival. Our proposed system comprises seven modules, as detailed below:

1. SSWM network design.

2. Vehicle route scheduling and monitoring phase.

3. Smart bin management and monitoring.

4. Workforce management with work schedules and attendance.

5. Staff workload management.

6. SWM command control center with grievance management for inquiry by the public.

7. Centralised database for maintenance and prediction.

Collection of Daily Waste

In the existing system, daily waste from households is collected through garbage vans by municipal corporation employees. This is done by making manual sounds, either by vocal or by blowing a whistle, which is sometimes inaudible or produces noise pollution. In the proposed system, we intend to address this issue by developing a mobile application on both the client and server side [13]. The mobile set-top devices will be activated automatically when the garbage truck is within 3 kilometres of the client's house (*i.e.*, the client-side mobile set-top box will emit an alarm sound to alert them that the garbage truck is approaching).On hearing this, the inmates of the households can keep their garbage ready for disposal. This prevents the garbage from being scattered by animals like stray dogs, birds, *etc*. This act will help to reduce the noise pollution depicted in Fig. (2).

Fig. (2). Daily waste collection.

Constrain-Based Collection

A provision has been added in the mobile application to dispose of loads of garbage immediately if there is an accumulation of loads of garbage that needs to be disposed of during special occasions in the home. In such cases, the mobile application will sense the signal from nearby garbage trucks and intimate it to the driver, either by making a call or sending a message to collect the garbage immediately. There should be provisions to register for garbage collection through a central monitoring system on such occasions. In such cases, the nearby garbage trucks are monitored and information about the client is sent to the driver and *vice-versa*, as like in cab booking systems in India (OLA, Umber taxi services). The trucks are constantly monitored to see if they reach the client's place on time by the GPS devices [14].

There are many densely populated and pilgrimage tourist destinations where there is a possibility of the generation of dangerous waste resources that need to be well organized in order to decrease the inappropriate consumption of valuable properties like human effort, time, and cost shown in Fig. (3). In our approach, we divided the overall system of waste detection into four components. *Via* the Smart Trash System, Smart Vehicle System, Local Base Station, and Smart Monitoring and Control Hut. All these sub-systems work intelligently and in coordination to automate the waste management in the Smart Bin. In this manner, waste can be disposed of as needed, rather than having to keep a constant eye on the waste bins manually.

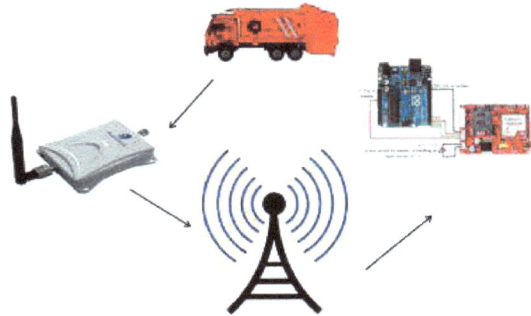

Fig. (3). Working of the signal boosting mechanism in mobile phones.

Smart Bin Design

Compaction

Every single clever bin is armed with ultrasonic sensors that measure the level of trash in the wastebasket. With its 240L wheelie bins for easy and safe trash removal, with the built-in safety sensor, detects motion and stops waste compaction. With incessant use, the baskets get filled up, progressively increasing their level. Every time the trash crosses a specific level, the sensors sense the data shown in Fig. (4). This data is promoted to analyse the levels of bin using Wi-Fi. The data is further used for the processes of analysis and predictive modelling. The solar-based Smart Bin is compatible with the standard 120L when it detects a hand. The bin's temperature sensor triggers an automatic compacting response to extinguish fire. All exterior access points are fitted with unique screws that protect against theft and damage.

- It is a completely automated process.
- It also helps monitor if vehicles (like BRC) are dumping the garbage at defined landfills or not.
- Compact Waste: Automatically compacts garbage to hold up to eight times the volume of standard trash cans.

Telecommunication

Communication information is collected in real time through wireless transmission to a website or app. 2G and 3G telecommunication modules are available through WCDMA and GSM modules.

ADVERTISEMENTS & CUSTOMIZABLE DESIGN

The entire solar waste bin can be wrapped and printed for marketing purposes, such as promoting new initiatives and encouraging recycling. What's more, Clean CUBE solar-powered trash cans can be equipped with LED backlights or LCD video screens to generate additional revenue through advertising.

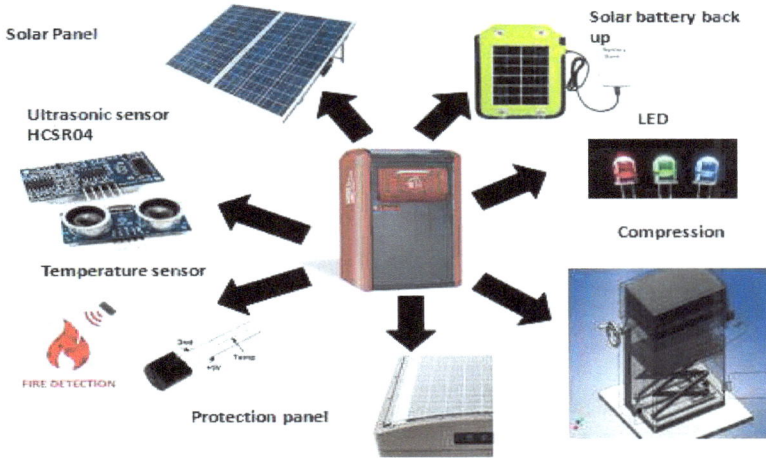

Fig. (4). Design of the smart bin.

LOCKING MECHANISM

All exterior access points are fitted with unique screws that protect against theft and damage, as shown in Fig. (**5**).

Fig. (5). Solar based smart bin locking system.

SOLAR COMPACTOR BENEFITS

- High Capacity – 600 litres – When full, automatic compaction allows for 5x the capacity of standard 120L wheelie bins.
- The flagship hopper design keeps waste contained while deterring pest access and preventing waste overflow and wind-blown litter.
- It will reduce street bin collections by an average 86%.
- Customization Options: Wraps, Stickers, Side Panels, Ashtray, Security Shield, Foot Pedal.

○ An internal compactor increases the litter capacity by 6–8 times that of a normal street bin.

○ It sends an email and text when the bin is 85% full.

○ Management and operatives can view, on desktops and smart phones, the real-time fill levels of all bins, eliminating "milk round" type collection rounds.

○ It eliminates overflowing bins.

○ It eliminates weekend collections.

○ It eliminates birds and vermin scattering rubbish across streets, parks and beaches.

○ Compostable Friendly: Public space compostable collection made with timely collection & enclosed design.

○ Monitor and report station fullness remotely.

○ No electricity is necessary for fullness level sensing or communication with the Smart Bin.

○ The Smart Bin Security System consists of two components: The Security Management Module that enables operations to set up and manage security events; and physical security plates that prevent access to the stations.

DETECTION OF HAND AND FIRE

The built-in safety sensor detects motion and stops waste compaction when it detects a hand. The bin's temperature sensor triggers an automatic compacting response to extinguish the fire shown in Figs. (6 and 7).

Fig. (6). SSWM hand detection.

FIRE DETECTION

Fig. (7). Fire detection in SSWM.

SMART BIN PROCESS AND MONITORING

Every smart bin is equipped with ultrasonic sensors which measure the level of the dustbin being filled up. The container is divided into three levels of garbage that is collected in it. With its continuous use, the levels get filled up gradually with time. Every time the garbage crosses a level, the sensors receive the data of the filled level. This data is further sent to the garbage analyser as an instant message using the GSM module. Every message that is received at the garbage analyser end is being saved as data, which is further used for the process of analysis and predictive modelling. The data received in real time is used by the application interface for a better view of the filled level. The data received is saved in the database, keeping all its attributes intact, such as time and date. A history of data collected over months is used by the department of data analysis for prediction and report making. Clean City Networks, or CCN, is the leading waste management platform and the glue that binds all our solutions together. CCN provides the monitoring environment, smart dashboard, analytics, and control centre in one comprehensive and simple package. Web-based and cloud-hosted, CCN is available anywhere you have a modern browser and an internet connection. It gives you total control and insight into your waste management operations and has proven cost-reduction benefits in all sectors of your operation.

DETAILS OF THE SMART SENSOR

Ardunio Uno Board

The Ardunio Uno is a microcontroller board based on the ATmega328P (datasheet). It has 14 digital input/output pins (of which 6 can be used as PWM outputs), 6 analogue inputs, a 16 MHz quartz crystal, a USB connection, a power jack, an ICSP header, and a reset button. As shown in Fig. (**8**), it contains everything needed to support the microcontroller; simply connect it to a computer with a USB cable or power it with an AC-to-DC adapter or battery to get started. Fig. (**9**) shows the Ardunio Uno board connected to an ultrasonic sensor, a thermal sensor, a compression box, a solar panel, an LED light, and a power backup battery.

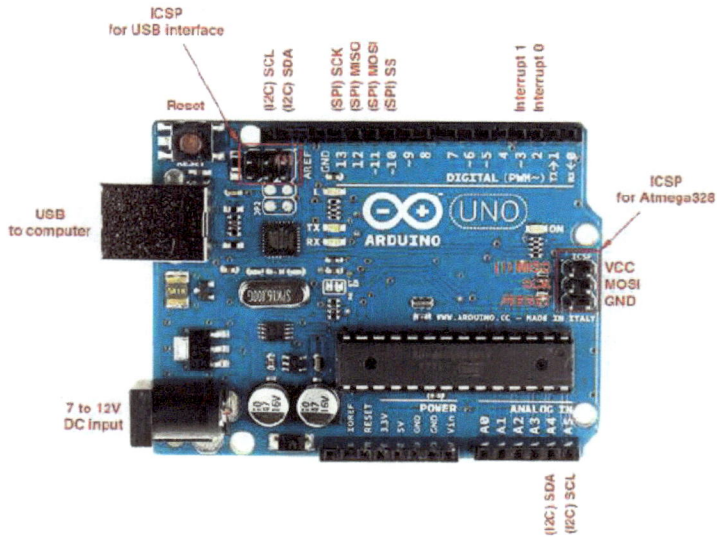

Fig. (8). Ardunio board.

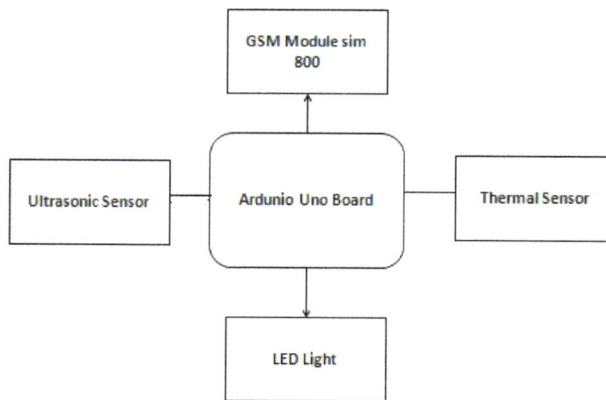

Fig. (9). Connections from the Ardunio board.

GSM Module SIM 800

This GSM Modem can accept any GSM network operator SIM card and act just like a mobile phone with its own unique phone number. Advantage of using this modem will be that you can use its RS232 port to communicate and develop embedded applications. Applications like SMS Control, data transfer, remote control and logging can be developed easily.GSM/GPRS MODEM is a class of wireless MODEM devices that are designed for communication of a computer with the GSM and GPRS network. It requires a SIM (Subscriber Identity Module) card just like mobile phones to activate communication with the network. Also they have IMEI (International Mobile Equipment Identity) number similar to mobile phones for their identification. A GSM/GPRS MODEM can perform the following operations:

1. Receive, send or delete SMS messages in a SIM.

2. Read, add, search phonebook entries of the SIM.

3. Make, receive, or reject a voice call.

The MODEM needs AT commands for interacting with the processor or controller, which are communicated through serial communication. These commands are sent by the controller/processor. The MODEM sends back a result after it receives a command. Different AT commands supported by the MODEM can be sent by the processor/controller/computer to interact with the GSM and GPRS cellular networks.

Ultrasonic Sensor

An ultrasonic sensor is a device that can measure the distance to an object by using sound waves. It measures distance by sending out a sound wave at a specific frequency and listening for that sound wave to bounce back. By recording the elapsed time between the sound wave being generated and the sound wave bouncing back, it is possible to calculate the distance between the sonar sensor and the object shown in Fig. (**10**).

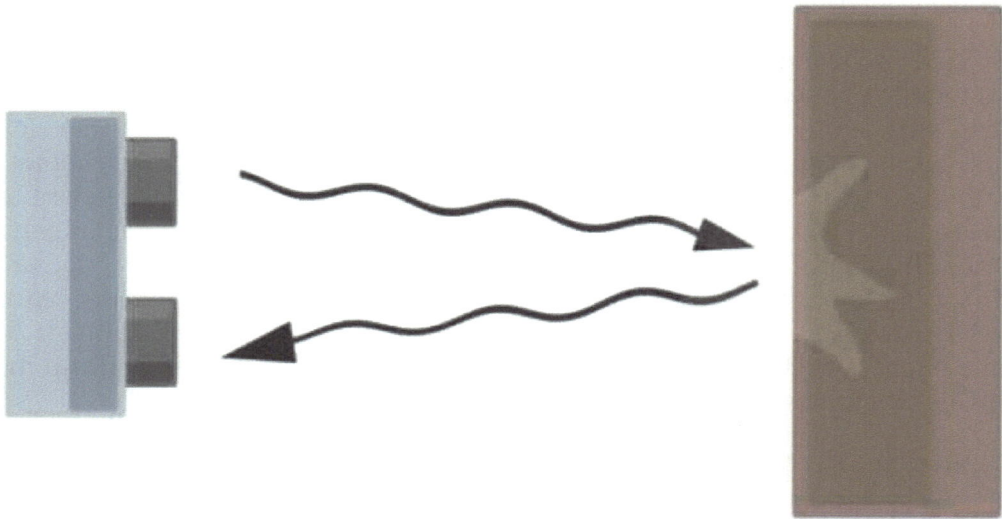

Fig. (10). Ultrasonic sensor.

Thermal Sensor

A thermostat is a contact type electro-mechanical temperature sensor or switch that basically consists of two different metals, such as nickel, copper, tungsten, or aluminum, *etc.*, that are bonded together to form a bi-metallic strip. The different linear expansion rates of the two dissimilar metals produce a mechanical bending movement when the strip is subjected to heat.

SMART BIN WORKING PRINCIPLES (SSWM)

The smart bin comprises of an Ardunio Uno microcontroller, sensors (ultrasonic and thermal sensors), a GSM module, a compaction unit, and an LED display. Trigger and Echo are the two pins of ultrasonic sensors used for calculating the time duration of the echo and the distance of the object by generating sound waves. The trigger pin sends eight 40 KHz sound waves, which is a high-to-low signal on the microcontroller. When the echo of the sound waves echoes back to the sensor, the ECHO pin turns high. The C3 pin of the ArduinoUno board is connected to the ECHO pin and is continuously monitored for detectstatus. The time period for the sound wave to travel back to the sensor is calculated by timer1 of the microcontroller. An ultrasonic sensor measures the distance (D) in cm, and the time taken (t) by the sound wave to echo back to the receiver is calculated in seconds, where v is the velocity of the sound wave. The distance of any solid waste material present in the bin is computed by the microcontroller. The central waste office receives a text message about the garbage level, which is detected by the GSM module. Each smart bin has a unique SIM card and a unique number, which acts as a unique ID. Large bins serve as the focal point of garbage collection in each locality. From these central bins, the garbage collection team collects the garbage. For this, central bins are converted into smart bins by applying a model of hardware. For this, dustbins are divided into three different levels according to the amount of garbage they hold. All bins in society send their level as a text message to a data warehouse. The central office candetect the level by sitting in their office in real time. This makes the officers send their workers to empty the bins. The hardware used is the basic electronic component and is also low cost. Irrespective of the size and height, hardware can be implemented in any bin and it is portable.

Smart Trash Bin Monitoring and Real-Time Analysis Service

The message service gives information about every level of the dustbin. An Excel sheet is formed by connecting this sheet with text files. An Excel sheet is used to show the filled level of every container. A text file is formed by SMS received from the GSM module. Then this text file was connected to the excel sheets. The vehicle is used to unload the bin. The vehicle's work is based on the level of the bin.

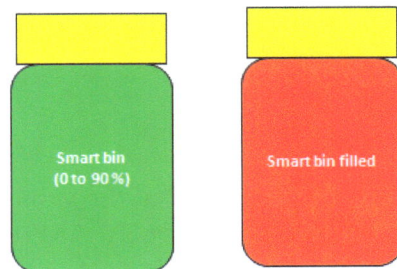

Fig. (11). Different levels of smart bin.

A real-time report is formed by the updated value of the dustbin level. Various excel functions like IFERROR, LARGE, INDEX, IF, COUNTIF, and ROW are used for the updated values of the excel sheet. Using charts in Excel, a widget is developed. Three level indications are made by the doughnut chart. The pointer, which moves accordingly to the level of the dustbin, is controlled by a pie chart in real time. A real-time dashboard is created by the excel application with a time series graph, which is used to obtain information about the current trend as well as the historical trend of each and every bin shown in Fig. (**11**).

AUTOMATIC VEHICLE ROUTE SCHEDULING AND MONITORING

Each department/concern person can track the route taken by the vehicle during the whole day with the reports on delays and time status of that vehicle. Vehicle Movement Tracking and Status can be checked on Centralized Computer Software and also through Mobile Apps. An alert sound is received in the mobile set-top box when the vehicle reaches the concerned household. Fig. (**12**) depicts the application dashboard.

Fig. (12). The outlook for SSWM.

The sample map view in Fig. (**13**) shows that the vehicle was scheduled to reach Point A at 6:00 AM but actually reached at 6:10 AM (a delay of 10 minutes) and collected a total of 3 garbage buckets from Point-A and Point-B.

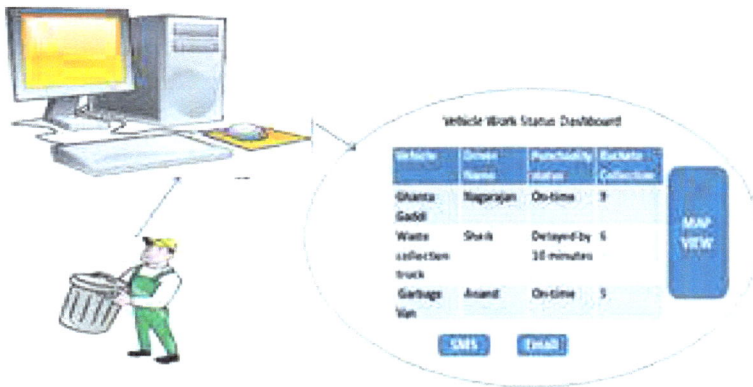

Fig. (13). Work schedule.

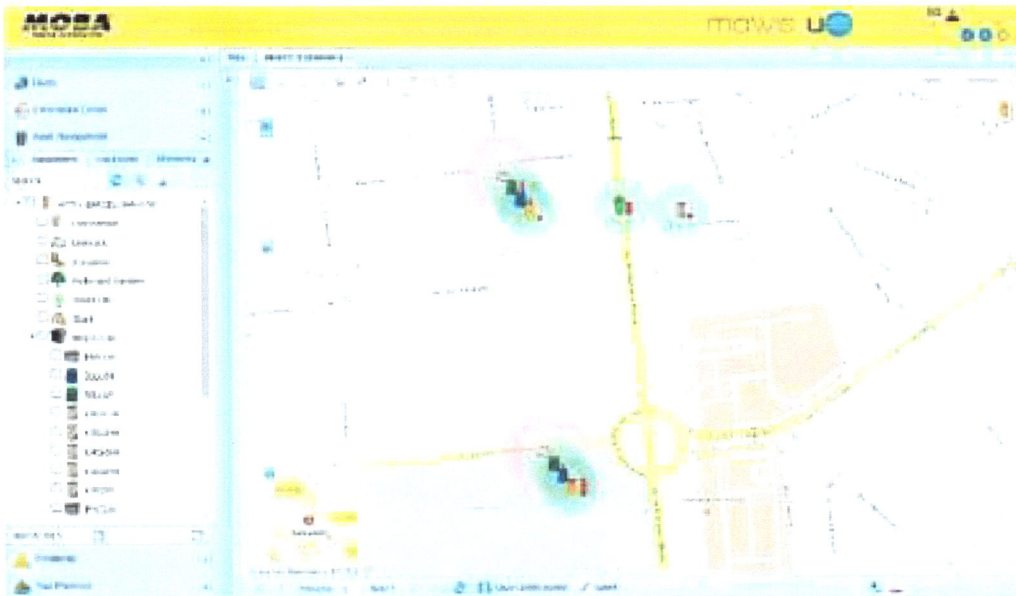

Fig. (14). Map work for vehicle route status.

WORKFORCE MANAGEMENT WITH WORK SCHEDULES

Online Work Allocation to the Staff

This will help to track their attendance and the quantity of work done by each person by providing information on the total number of buckets unloaded by them on the truck shown in Fig. (**14**).

Amount of Work Done by the Staff/Work Load Calculation

The garbage collector person will show I-button tags on the I-button readers installed on trucks while unloading the garbage bins. This will help to track their attendance and the quantity of work done by each person by providing information on the total number of buckets unloaded by them on the truck.

SSWM COMMAND CONTROL CENTER WITH GRIEVANCE MANAGEMENT FOR PUBLIC ENQUIRY

Providing a Cloud-Based Citizen Grievance Redressed Management System, website and mobile application. To establish a single point of contact for responding to citizen grievances received *via* any of the communication channels, such as an online portal, mobile application, e-mail, manually on paper, social media, or call centre number. This grievance system will act as a centralised citizen's grievance redressal system for all the grievances shown in Figs. (**15** and **16**).

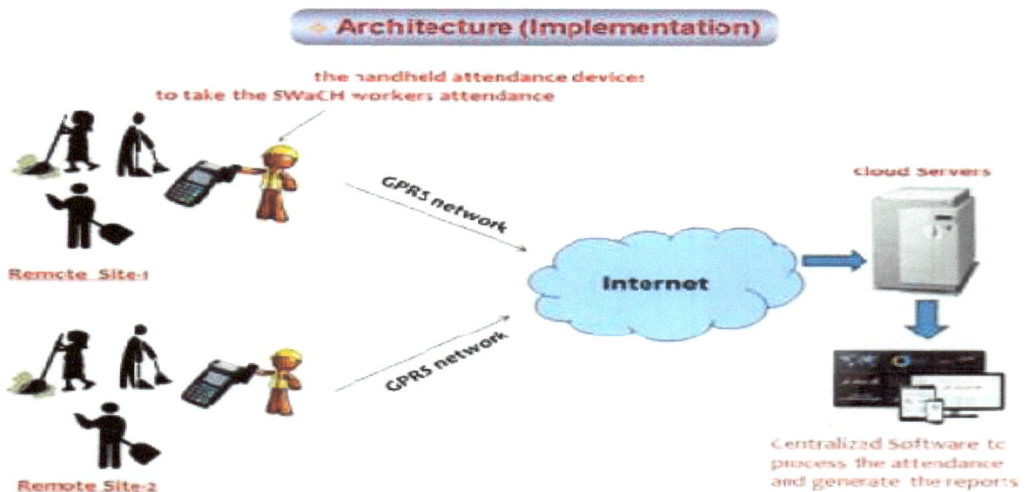

Fig. (15). Overview of architecture implementation.

Fig. (16). Centralized grievance system.

RESULTS AND OUTCOMES

Fig. (17). Ardunio board connection.

Fig. (18). LED light for showing the level of garbage.

CONCLUSION AND FUTURE WORK

Developing and developed countries are facing high pressure to meet objectives to collect and recycle waste. However, waste collection and recycling waste costs pose major problems at the municipal level for the waste management system. The empirical literature on waste management has paid much more attention to demand-side aspects (reduction and discouragement of land disposal and promotion of recycling and recovery) than to supply-side issues such as waste management costs. This paper addresses the gap in this research field by estimating the cost function of providing waste collection and recycling services. The results of the proposed model may lead to more efficient waste management shown in Figs. (**17** and **18**). In the future, the datasets stored for data analysis will be used for data analysis.

CONSENT FOR PUBLICATION

Not applicable.

CONFLICT OF INTEREST

The author declares no conflict of interest, financial or otherwise.

ACKNOWLEDGEMENT

Declared none.

REFERENCES

[1] S.S. Hahn, J. Kim, and S. Lee, "To collect or not to collect: Just-intime garbage collection for high-performance ssds with long lifetimes", In: *in Proceedings of the 52Nd Annual Design Automation Conference*ser. DAC '15. New York, NY, USA: ACM, , 2015, pp. 191:1-191:6.
[http://dx.doi.org/10.1145/2744769.2744918]

[2] McEwan, and M. Z. Komsul, *Pre-emptive garbage collection for ssd raid," in Euro micro Conference on Digital System Design (DSD),* pp. 356-363, 2016.

[3] Lindawati, C. Wang. C. Cui, and N. Hari, *Feasibility Analysis on Collaborative Platform for Delivery Fulfillment in Smart City in SmartCity,* pp. 147-152, 2015.

[4] D. Puiu, P. Barnaghi, R. Tonjes, D. Kumper, M.I. Ali, A. Mileo, J. Xavier Parreira, M. Fischer, S. Kolozali, N. Farajidavar, F. Gao, T. Iggena, T-L. Pham, C-S. Nechifor, D. Puschmann, and J. Fernandes, "CityPulse: Large Scale Data Analytics Framework for Smart Cities", *IEEE Access,* vol. 4, pp. 1086-1108, 2016.
[http://dx.doi.org/10.1109/ACCESS.2016.2541999]

[5] S. Shahrestani, *"Assistive IoT: Deployment Scenarios and Challenges," in Internet of Things and Smart Environments.* Springer, 2017, pp. 75-95.
[http://dx.doi.org/10.1007/978-3-319-60164-9_5]

[6] B. Li, and P.D. Franzon, "Machine learning in physical design", *Proc. of the IEEE,* pp. 147-150, 2019.

[7] PravinChopade, *Saad M Khan, David Edwards, and Alina von Davier.* Machine Learning for Efficient Assessment and Prediction of Human Performance in Collaborative Learning Environments: USA, 2018.

[8] Farhad Ahamed, and Farnaz Farid, *Applying Internet of Things and Machine-Learning for Personalized Healthcare: Issues and Challenges International Conference on Machine Learning and Data Engineering (ICMLDE).,* 2018.

[9] Amrutha, Kavyashree, and Pooja, "IoT Based Waste Management Using Smart Dustbin", *International Journal of Science,Engineering and Technology.,* 2017.

[10] shah Rushabh, "IoT Based Smart Bin", *International Research Journal of Engineering and Technology.,* vol. 04, no. 09, 2017.

[11] Minthu Ram Chiary, "SripathiSaiCharan,AbdulRashath,Dhikhi, "Dustbin Management System Using Iot"", *International Journal of Pure and Applied Mathematics. Volume,* vol. 115, no. 8, 2017.

[12] S. Revathy, "Iot Based Smart Bin Monitoring Using Sensor And Gsm For Smart Cities", *International Journal of Research in Computer Science. ,* pp. 2339-3828, 2019.

[13] AbhishekDev, "IoT Based Snart Garbage Detection System", *International Research Journal of in Engineering and Technology (IRJET).,* vol. 03, 2016.

[14] Prabu Parkash, "IoT Based Waste Management for Smart City", *International Journal of Innovative Research inComputer and Communication Engineering (An ISO 3297: 2007 Certified Organization).,* vol. 4, no. 2, 2016.

CHAPTER 4

Remediation Approaches for the Degradation of Textile Dye Effluents as Sustaining Environment

S. Lakshmana Prabu[1,*], R. Thirumurugan[2], M. Rengasamy[3] and G. Venkatesan[4]

[1] *Department of Pharmaceutical Technology, University College of Engineering (BIT Campus), Anna University, Tiruchirappalli, India*

[2] *College of Agriculture and Life Sciences, North Carolina State University, Kannapolis, NC, USA*

[3] *Department of Petrochemical Technology, University College of Engineering (BIT Campus), Anna University, Tiruchirappalli 620 024, India*

[4] *Department of Civil Engineering, University College of Engineering (BIT Campus), Anna University, Tiruchirappalli 620 024, India*

Abstract: Water has been considered one of the most valuable substances on earth for almost entire living organisms, from the largest mammal to the smallest microorganism. In addition, water is essential for the healthy life of human beings, animals, plants, *etc.* due to rapid, swift, and advanced industrialization, polluted water is discharged from different industries on many occasions. Among the different industrial pollutants, textile dyes and their effluents are the most predominant pollutants. Nearly 100,000 synthetic dyes are produced, and about one million tons of dyes are utilized for various dying purposes. About 10% of the dyes are unconfined into natural resources and the environment as waste, which spoils the aesthetic nature of the environment. These colored dyes are carcinogenic or mutagenic. These colored dyes are very fine particles in nature, and their concentrations of about 1 ppm are visible. These discharged color dyes cause grave intimidations with numerous problems; hence, these discharged color dyes as industrial waste have been considered as a major problem in the wastewater treatment process. In this chapter, various remediation techniques for the degradation of textile dyes effluents are discussed to maintain the sustainability of the environment.

Keywords: Biological methods, Effluent treatment techniques, Electrochemical Methods, Physical-Chemical methods, Phytoremediation Green Nanotechnology, Textile dye, Azo dye.

---

* **Corresponding author S. Lakshmana Prabu**: Department of Pharmaceutical Technology, University College of Engineering (BIT Campus), Anna University, Tiruchirappalli, India; Email: slaxmanvel@gmail.com

G. Venkatesan, S. Lakshmana Prabu and M. Rengasamy (Eds.)

INTRODUCTION

The foremost important substance on the earth is water, which has been fundamentally required for the survival of every living organism from the biggest mammal to the smallest organisms. In simple words, without water, there will be no life on earth. Animals, plants, and organisms utilize water for their healthy growth as well as for their production by aerobic respiration; whereas without water there is no aerobic respiration. Commonly the living organism creates ATP as a source of energy for their life through the aerobic respiration process. Most of the organisms are made up of a minimum of 50% however few of the organisms are made with 95% of water. Water must be clean for humans to have a healthy life also it should be free from chemicals and microorganisms (https://sciencing.com/about-6384365-water-important-life-earth-.html) [1].

Diverse water resources are:

1. Springs

2. River water

3. Surface water

4. Rock holes and rock catchment areas

5. Excavated dams

6. Rainwater

7. Bores and wells

8. Artesian bores (http://www.health.gov.au/internet/publications/publishing.nsf/ Content/ohp-enhealth-manual-atsi-cnt-l~ohp-enhealth-manual-atsi-cnt-l-ch6~ohp-enhealth-manual-atsi-cnt-l-ch6.1) [2].

Derivation of Synthetic Dye

From ancient times, different colorants have been utilized by human beings from their routine life practices like painting, cloth coloring, dyeing, *etc.*, Walls were colored by using dyes during 15000-9000 BC itself (Altamira caves in Spain). Until the 17th-century insects, plants and mushrooms were used as a natural dye source especially in textile dying (Shah 2018) [3]. In general, these natural dyes are made with aromatic compounds; while these dyes require several steps in the dyeing process, which is a main drawback of the same.

In 1856, W.H. Perkin has tried to synthesis quinine, unfortunately, the outcome of that particular research produced a blue color substance that had respectable dyeing character. Subsequently blue color has turned into violet aniline and purple tyrant color (Aksu and Karabayir 2008) [4].

Rapid industrial advancement and development made a new demand in the dyeing technique which created a new gateway in the dyeing industry to produce different synthetic dyes. To encounter the present need, about ten thousand types of dyes are manufactured; also almost 1,000,000 tons of dyes are produced exclusively for textile industry purposes (Robinson *et al.,* 2001) [5].

Classification of Dyes

Dyes are molecules having conjugated double bonds in delocalized electronic systems; chromophore and auxochrome are the major groups in the dyes. In general, a chromophore is an electron-withdrawing group of atoms that have a major role in the dye color. Some vital chromophore groups in the dyes are -N = N-, -C = N-, -C = O, -C = C-, -NO and $-NO_2$. But auxochrome is an electron donor substituent group and responsible for the electronic system energy modification, which increases the chromophore color. Some vital auxochrome groups in the dyes are - NH_2, -NR_2, NHR, - SO_3H, -OH, -OCH_3, and –COOH (Aksu and Tezer 2005; Alhassani *et al.,* 2007) [6, 7].

The chemical molecular structure of the dyes has an important role in the dye's color against the decline of color when exposed to light and water. Textile industry dyes are categorized into anthraquinone, basic, azo, diazo, cationic, anionic, and nonionic based on their chemical structure. However the dyes consist of anthraquinoid, azo aromatic, and indigoid structure which allow strong π-π transitions through high extinction co-efficient in UV visible area; but the azo aromatic dye is the more widespread between these three dyes (Palanivelan *et al.,* 2013) [8].

Azo dyes have one or more nitrogen–nitrogen double bonds (–N==N–); further classified into mono-azo dyes, diazo dyes, tri-azo dyes, and poly-azo dyes (more than three azo groups) based on the number of azo groups. The major merits of these azo dyes are ease of synthesis and cost-effectiveness; demerits are such as it can't be simply besmirched by aerobic and anaerobic bacteria (Shivangi 2012) [9].

The classification of dyes is shown in Table **1**.

Table 1. Classification of dyes.

Natural Dyes	Synthetic Dyes
From Plants – Madder (Madder Root)	*Azo Dyes* – Acidic, Basic, Reactive, Disperse, Sulfur and vat
From Animal – Tyrian purple (sea snails)	*Non-Azo dyes*

Significance of Effluent Treatment in Textile Dye Industry

Modernization, advancement, and rapid industrial growth in the urban areas have made a major path in water contamination. Throughout history, on many occasions, living organisms were died due to the discharge of untreated polluted water from different industries (http://www.health.gov.au/internet/publications/ publishing.nsf/Content/ohp-enhealth-manual-atsi-cnt-l~ohp-enhealth-manual-atsi-cnt-l-ch6~ohp-enhealth-manual-atsi-cnt-l-ch6.1) [2]. Among the different industries, textile industry untreated polluted water has been considered as a major issue. As assessment made by the World Bank demonstrated that among the different polluted water, polluted water from the textile industry is in the range from 17% to 20% (Holkar *et al.,* 2016; Rani *et al.,* 2013) [10, 11]. As earlier pointed out that about ten lakh tons of dyes are exploited for dyeing in the textile industries, in which about 10% of dyes are cleared as waste into different natural sources (Jadhav *et al.,* 2010) [12]. A survey result outlined that 1.6 million liters of water is used in a single day for the manufacturing of 8,000 kg of fabric material. In simple for 1 kg of textile fabric process, 200 liters of water is used (Khandare and Govindwar 2015) [13]. After treating 1000 kg of fabric material made with cotton, the composite discharge waste may have 30 to 50 ppm of suspended solids, 1000 to 1600 ppm of total solids, and 200 to 600 ppm of biochemical oxygen demand (BOD) in a volume of 50 to 160 m^3 (Kimmatkar *et al.,* 2017) [14].

Textile dye in small quantity can disturb the water quality.Its intense color can prevent sunlight penetration, being more stable to oxidation, pH alteration having a complex structure. It also decreases the level of dissolved oxygen, increases the chemical oxygen demand (COD) and BOD and is resistant to a breakdown which leads to depreciate the quality of water by pollution generation, eutrophication and distraction of life in aquatic system. These properties of azo dye produce hazardous and harmful metabolites in the aquatic system and it becomes toxic to aquatic living fish, mammals, as well as to public health (Mahdavi Talarposhti *et al.,* 2001) [15]. Azo dye concentration less than 1 ppm can be perceptible, significantly damaging the environment ecosystem as well as water system (Gupta *et al.,* 2003) [16].

Reaction among the synthetic dyes and other chemicals can produce non-biodegradable by-products. These textile effluents contain not only the azo dyes, addition, but also a complex mixture of other components like textile auxiliaries, chemicals, different salts of surfactants, heavy metals, mineral oils, *etc.* (Hassan *et al.,* 2009) [17]. Various reports revealed that different inorganic and organic compounds such as naphthol, nitrates, sulfur, acetic acid, enzymes, chromium compounds, soaps, heavy metals including lead, copper, nickel, mercury, cadmium, arsenic, cobalt, and some other auxiliary chemicals are present in the textile effluent. Other organic compounds include strain remover (chlorinated compound), dye fixing agent (formaldehyde), softeners (hydrocarbon-based compounds), and non-biodegradable dyeing chemicals. These inorganic and organic compounds can cause toxic as well as carcinogenic effects on living organisms.

Earlier reports established that about 100,000 types of dyes are produced and used in textile industries, among these dyes, 72 dyes have been considered toxic. In addition among these 72 dyes, about 30 dyes are unable to remove by effluent treatment technique (Chen and Burns 2006) [18]. From effluent removal of colored dye, substances are a very critical and vital one, hence textile industries effluents are recognized and considered as the main pollutant when compared to other pollutants (Bharathiraja *et al.,* 2018) [19].

Impacts on Public Health

The maximum of the azo dyes has been considered carcinogenic and mutagenic (Pinherio *et al.,* 2004) [20].

Mechanisms for its carcinogenic nature are:

- Cleavage of azo bond and release of aromatic amines from azo dyes, which covalently binds with DNA cause carcinogenic effect.
- Without azo reduction release of the free aromatic group from azo dyes to form metabolically oxidized form.
- Azo bond oxidation directly with electrophilic diazonium salts of azo dyes (Bharathiraja *et al.,* 2018) [19].

The mutagenic nature of azo dyes is due to:

• Aromatic amines like methylamines, chloro aniline, *etc.*, (Shivangi 2012) [9].

The occurrence of degraded products of these azo dyes can cause several illnesses in humans including hemorrhage, nausea, vomiting, allergies, skin infections, ulceration, *etc.* Also these azo dyes cause kidney, liver, brain, central nervous system, and reproductive system damage. Also, chromosomal deformities in mammalian cells and different cancers such as bladder, spleen, liver were reported (Chung and Chen 2009; Ali 2010; Pooja *et al.,* 2019) [21 - 23].

Dye Removal Techniques From Effluent

Effluent regulatory guidelines and disposal costs made that effluent treatment techniques have an important key role in the protection of various water resources as well as the environment. The main principle in the treatment of wastewater is to eliminate the pollutant instead of destroying it from the effluent.

The main objectives of the effluent wastewater treatment:

• To safeguard the environment.
• To protect socio-economic concerns.
• To provide good public health (Metcalf and Eddy 1991) [24].

Different effluent treatment techniques are shown in Table **2**.

Table 2. Different effluent treatment techniques.

S. No.	Method	Techniques
1	Physical	•Filtration • Adsorption • Coagulation/Flocculation • Reverse osmosis
2	Chemical	• Oxidation • Ozonation • Electrolysis
3	Biological	• Microorganism • Enzymes • Plants

In dye removal, each technique has its restrictions. Based on the nature of the effluent from the textile industries, an appropriate technique needs to be designed to produce an efficient effluent treatment. The various factors that govern the selection of an appropriate technique is:

- Pollutants/Contaminants limits in the effluent from the industry.
- Physical and chemical properties of the effluent.
- Generated solid waste nature after treatment.
- Cost/Economic feasibility.
- Environment (Lakshmana Prabu *et al.,* 2016) [25].

Physical-Chemical Effluent Treatment Techniques

Dye contaminated wastewater can be treated by different methods, which are (Ahmad *et al.,* 2015) [26]:

- Filtration (Membrane separation)
- Coagulation/Flocculation
- Ion exchange
- Advanced oxidation process
- Adsorption by activated carbon
- Fenton's reagent
- Ozonization
- Photochemical
- Destruction by electrochemical method
- Irradiation technique
- Coagulation by an electrokinetic method

Filtration (Membrane Separation)

In this filtration process, based on the pore size, solid contaminants from the effluent are separated. Each filtration aid has a definite arrangement at the molecular level and its membrane structure provides the desired pore size of the membrane to produce efficient separation of the solids. Recycling of specific contaminants from the effluent can also be treated by this technique. Based on the pore size and membrane nature, this technique is further classified into:

  I. Microfiltration -appropriate for the dyes having dye pigment.
 II. Ultrafiltration - removal of macromolecules and particles.
III. Nanofiltration - used to treat colored wastewater.
IV. Ion exchange –Used for removing both cationic and anionic dyes.
 V. Reverse osmosis – Elimination of chemical additives by degradation in the dye (Shah 2018) [3].

Type of Membranes

In the filtration method, either synthetic or natural membranes are utilized for the separation of solid contaminants.

- The natural membrane includes cellulose, wool, and rubber (polyisoprene).
- The synthetic membrane includes polyamide, polystyrene, and polytetrafluoroethylene (Teflon).
- The inorganic non-polymeric membrane includes zeolites, metal, ceramic and carbon.
- The hybrid membrane includes varied matrix membranes of both organic and inorganic compounds.
- Bipolar membranes include diverse ionic charge material complex (https://www. asahi-kasei.co.jp/membrane/microza/en/kiso/kiso_1.html; http://www.separationprocesses.com/Membrane/MT_Chp03.htm; Mohan *et al.,* 2014) [27 - 29].

Coagulation/Flocculation

Coagulation/Flocculation is a physicochemical process; in this process, coagulant/flocculant is added in the effluent which promotes the gathering of fine particles to produce a coagulation/flocculation and this can be easily separated through the filtration technique. Also, it helps attain the maximum COD removal from the effluent.

There are two types of coagulants. They are:

1. Primary coagulants – Electrical charges of the suspended solid pollutants in water is neutralized.

2. Coagulant aids – Slow settling flocs density is increased by impacting toughness.

Chemicals used as coagulants are either metallic inorganic salts or polymers. Polymers are classified into:

- Cationic
- Anionic
- Non-ionic

Coagulation is an aggregation of fine particles with a specific type that forms compact aggregates and flocculation is an aggregation of fine particles that forms loose or open aggregates.

Coagulants commonly used in this technique are:

- Ferrous sulfate
- Ferric chlorosulfate
- Ferric chloride
- Alum (aluminum sulfate)

The chemical reaction involved in the coagulation process by a common coagulant (alum) is given as:

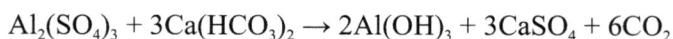

$$Al_2(SO_4)_3 + 3Ca(HCO_3)_2 \rightarrow 2Al(OH)_3 + 3CaSO_4 + 6CO_2$$

Mechanisms involved in Coagulation/Flocculation are:

a. Sorption (amino group protonation).
b. Electrostatic attraction.
c. Bridging (interrelated with high molecular weight polymer).

(Hassan *et al.,* 2009; http://www.profilt.net/en/water-treatment-principles/; http://www.columbia.edu/~ps24/PDFs/Principles%20of%20Flocculation%20Dispersion%20Selective%20Flocculation.pdf; Sabur *et al.,* 2012; Ukiwe *et al.,* 2014) [30 - 32].

A schematic diagram for coagulation-flocculation (Mazille and Spuhler-https://sswm.info/sswm-university-course/module-6-disaster-situations-planning-and-preparedness/further-resources-0/coagulation-flocculation) is shown in Fig. (1) [33].

Ion Exchange

In this method, undesirable ions are removed with the help of ion exchange resin beds (cationic/anionic) and replaced with less objectionable ones. By incorporating appropriate resin beds, selective ions can be removed completely from the effluents.

It is an effective multipurpose process to eliminate most of the hazardous compounds from the effluent also to concentrate the effluent.

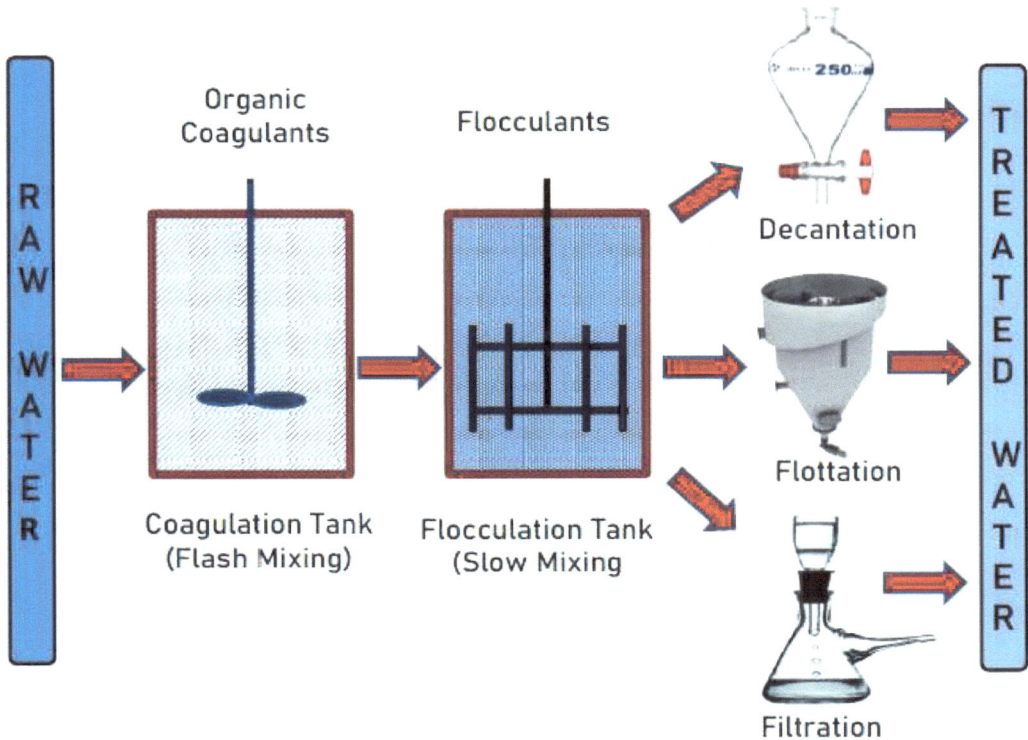

Fig. (1). Schematic diagram for coagulation-flocculation.

Ion exchangers commonly used in this technique are:

- H^+ (proton) and OH^- (hydroxide).
- Monovalent charged ions such as Na^+, K^+, and Cl^-.
- Divalent charged ions such as Ca^{2+} and Mg^{2+}.
- Polyatomic inorganic ions such as SO_4^{2-} and PO_4^{3-}.
- Amine functional groups in organic bases such as $-NR_2H^+$.
- The carboxylic group as function group in organic acids such as $-COO^-$ (https://textilelearner.blogspot.com/2012/05/base-exchange-ion-exchange-water.html; Parameswaran *et al.,* 2014; Wawrzkiewicz and Hubicki 2015) [34 - 36].

The Schematic diagram for ion exchange is shown in Fig. (**2**).

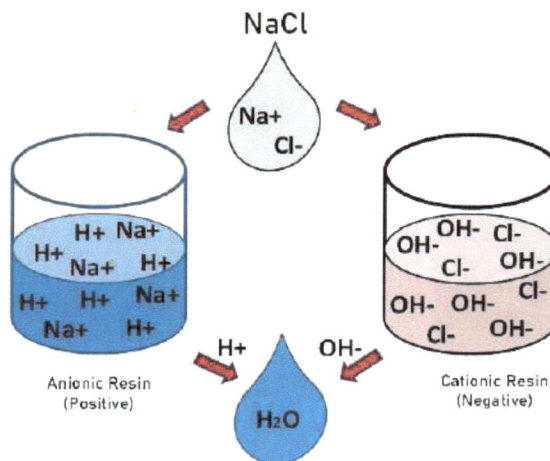

Fig. (2). The Schematic diagram for ion exchange.

Advanced Oxidation Process

The advanced oxidation process is a process with the combination of different chemical treatment methods which will create an influential method for hydroxyl radical. In this process, both inorganic and organic components in the effluent are oxidized (Pesoutova *et al.,* 2011) [37]. Fenton oxidation process is a very good example for this advanced oxidation process and the Fenton oxidation process (Huang *et al.,* 2017) is shown in Fig. (**3**) [38].

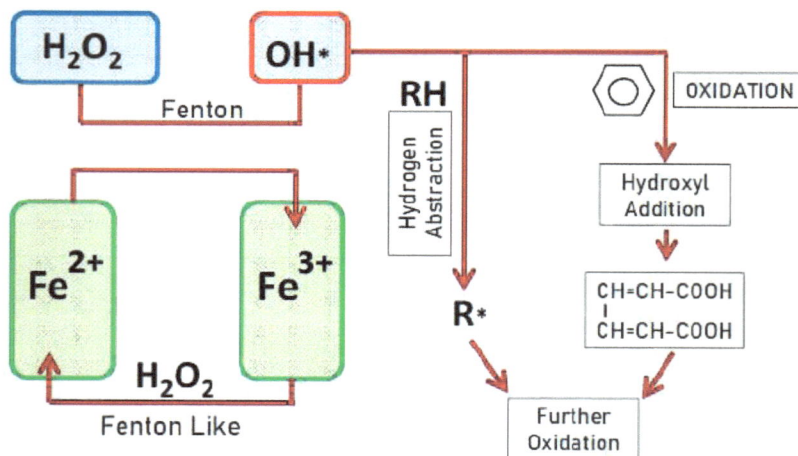

Fig. (3). Fenton oxidation process.

Different types of advanced oxidation processes are (Pera-Titus *et al.,* 2004) [39].

a. Fenton Process: This process is a simple process to generate OH radicals; this process generates OH radicals by the reaction among H_2O_2 to Fe^{2+} salts. The chemical reaction involved in this process is shown as:

$$H_2O_2 + Fe^{2+} \rightarrow OH^{\cdot} + OH^{-} + Fe^{3+}$$

b. Electro-Fenton Process: In this process, OH radicals are generated by the reduction of oxygen through an electrochemical method in the presence of ferrous ions.

c. Photo-Fenton Process: In this process, OH radicals are produced on the ferrous ions by UV radiation.

$$H_2O_2 + UV \rightarrow OH^{\cdot} + OH^{-}$$

Adsorption by Activated Carbon

In this method, effluent components are adsorbed by bonding on the surface of activated carbon through either physical or chemical. It is a simple, cost-effective, and efficient method to eliminate both inorganic and organic pollutants from the effluent.

Physical sorption (Van der Waals forces) or chemisorption (chemical bonding) is the basic sorption of pollutants. Activated carbon is an amorphous nature having various pore sizes (Pera-Titus *et al.,* 2004; Foo and Hameed 2010; Popuri and Guttikonda 2015) [39 - 41]. This adsorption process takes place between the adsorbents and pollutants in four phases. They are advective transport, film transfer, mass transfer, and intraparticle diffusion. The adsorption process (Patel 2018) is shown in Fig. (**4**) [42].

Ozonation

Oxidation property of the effluent components is the basic principle of this method and it has been preferably used to treat textile effluents. In this technique, the oxidation takes place by the chemical reaction between hydroxyl radicals of the effluent compounds preferably on the unsaturated bonds of chromophores (conjugated double bonds) of the dye and ozone. Ozone compounds in water gets decomposed with OH and produces hydroxyl radicals and oxidize the components either directly or indirectly (Shriram and Kanmani 2014) [43].

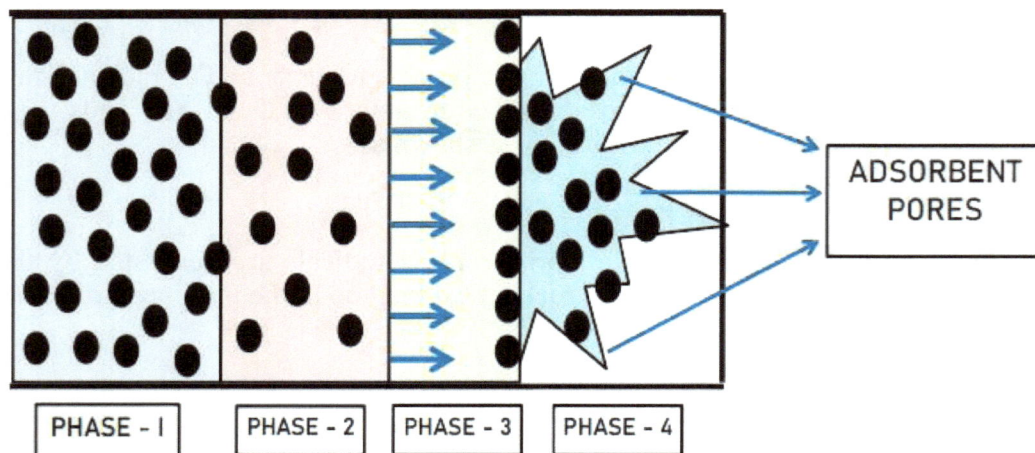

Fig. (4). Adsorption process between the pollutants and adsorbents.

Direct – Dissolved molecular ozone (O_3) react directly and oxidize the compounds.

Indirect – Radical species (HO, HO_2) react indirectly and oxidize the compounds.

Factors affecting the ozonation process are:

- Ozone dose
- pH of the medium
- Dye concentration
- UV radiation
- Temperature

Irradiation

Irradiation is an exposure of ionizing radiation process in textile effluent treatment. Gamma rays (electromagnetic radiation) have been used in high frequency with sufficient energy as a radiation source to produce ions by disentangling the electrons from molecules and atoms; the subsequent reaction of ions to form free radicals. On radiolysis of water molecules by these free radicals, subsequently these molecules undergo a chemical reaction with effluent dye components (azo dyes) and convert into amide which is further converted into ammonium compounds (Rahman Bhuiyan *et al.*, 2014; Parvin *et al.*, 2014) [44, 45].

Electrochemical Methods

Electrochemical methods are classified as:

1. Electrochemical oxidation

2. Electrocoagulation

3. Electro floatation

Electrochemical Oxidation

In this method, the pollutant compounds are changed into simple components like CO_2 and H_2O by either direct or indirect oxidation.

Direct oxidation - Anodic oxidation is the direct oxidation process that adsorbs the pollutant components on the anode surface and is oxidized by an anodic electron transfer reaction.

Indirect oxidation - In this method, potent oxidants such as chlorine, hydrogen peroxide, hypochlorite, ozone, and hydroxyl ions are generated on the surface of anode electrochemically in two stages.

First stage – At the anode surface, hydroxyl radical is generated by either acidic solution or alkaline solution.

Second stage – Higher oxide components are produced between the metal oxide and the adsorbed hydroxyl radical (Feng *et al.,* 2016; Radjenovic and Sedlak 2015; Martínez-Huitle and Andrade 2011) [46 - 48].

Electrocoagulation

Electrostatic charges of pollutants are neutralized followed by the formation of coagulation, which will assist to purify the effluent water. Simultaneous formations of electrocoagulation followed by electro flocculation are the basic principle of this technique.

In electrocoagulation, corresponding metal ions are produced with suitable electrodes such as an aluminum electrode for aluminum or ion electrodes for ion, which form a multi-charged polynuclear complex. This multi-charged polynuclear complex has enhanced adsorption property and assists in the formation of coagulation of the pollutants from the textile effluent. Immediately hydrogen gas is produced from the cathode; it will support the coagulated particles to float on the surface of the water. Metal salts such as $FeSO_4$, $Fe_2(SO_4)_3$, $FeCl_3$, and $Al_2(SO_4)_3$ are preferably used in this electrocoagulation process.

A typical electrocoagulation process (Canizares *et al.,* 2005) is shown in Fig. **(5)** [49].

The electrocoagulation process takes place in different stages. They are:

1. Electrolytic reactions at the electrode

2. Coagulation formation

3. Adsorption of pollutants on the surface of coagulation

4. Removal of pollutants by flotation or sedimentation

The electrocoagulation process depends on the following factors:

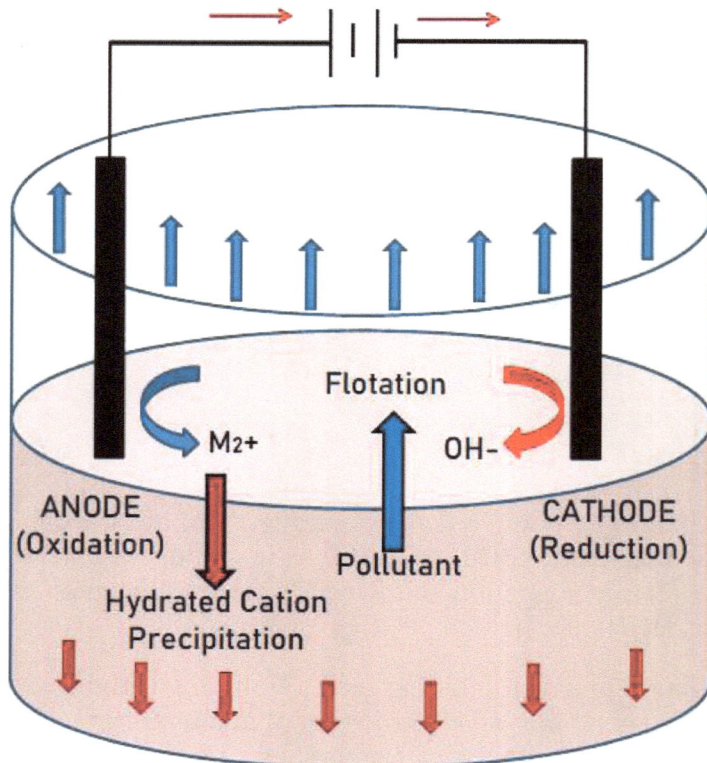

Fig. (5). Schematic diagram of an electrocoagulation process.

- Effect of electrode material
- Effect of inter-electrode distance
- Effect of electrolyte (NaCl) concentration
- Effect of current density and applied voltage
- Electrical energy and electrode consumption
- Effect of pH
- Effect of operating temperature
- Effect of electrocoagulation time
- Effect of concentration of dye

Electro Flotation

The subsequent process of electrocoagulation is Electro flotation. Small sized gas bubbles such as oxygen from the anode and hydrogen from the cathode are evolved by electrochemical reactions. These evolved gases will help the pollutants to float on the water surface.

Electro Flotation process takes place in different stages. They are:

1. Gas bubble generation

2. The interface between pollutants and gas bubble

3. Gas bubble adsorption on the pollutants surface

4. Rising of gas bubble along with pollutants on the surface of the water

Electro Flotation process depends on the following factors:

1. Gas bubble diameter and zeta potential

2. Surface tension of the solution

3. Particle size of pollutant

4. Pollutant concentration

5. The residence time of water in the electrolytic cell

6. Temperature

7. pH (Arulmurugan *et al.,* 2007; Butler *et al.,* 2011; Ukiwe *et al.,* 2014; Naje *et al.,* 2015; Abbas and Ali 2018) [50 - 54].

Biological Methods

The above said Physico-chemical methods have their restrictions, higher cost, less efficiency, generation of hazardous chemicals and toxic compounds, *etc.* further secondary waste components generations which are needed to be tackled further.

Newly biological methods have received abundant consideration in treating dye effluent due to its easy process, cost, rate of synthesis, free from toxic chemicals, can be used to eliminate various categories of dye effluents. Different biological sources such as fungi, bacteria, microorganisms, algae, and plants have been utilized in biological methods. Among the different biological sources, plants (Phytoremediation) are the best candidates due to their potential, economical, free from toxic chemical utilization, availability, protective, phytoconstituents can provide a superior plant form, *etc.* (Kore *et al.*, 2017 and Tang *et al.*, 2022) [55, 56].

Phytoremediation

Phytoremediation - Plant properties are used to remove the pollutants from the environment. The term Phyto is derived from the Greek word "Phyto" meaning "Plant", whereas the term Remediation is derived from the Latin word "Remediation" meaning "Cure". Recently this approach has become a developing technology and gathered much interest among the researchers for treating dye effluent due to its positive results and cost-effectiveness. It is an eco-friendly method, which can remove several categories of pollutants such as metal hydrocarbons, pesticides, chlorinated solvents, *etc.* (Macek *et al.,* 2000; Susarla *et al.,* 2002; Xia *et al.,* 2003) [57 - 59].

In this approach, pollutants from the effluent are removed by:

1. Assimilation

2. Degradation

3. Metabolization

4. Pollutant detoxification (Vasanthy *et al.,* 2011) [60].

Mechanisms of the Phytoremediation process are:

1. Uptake of contaminants, aggregation, and metabolization in plant tissue.

2. Transport of volatile organic hydrocarbons through leaves.

3. Motivation of microbiological activity by exudate release and biochemically transfer near to the roots for metabolism.

4. Enhanced mineralization at the surface of root-soil that is predicted by mycorrhizal fungi and microbial consortia (Schnoor *et al.,* 1995) [61].

Classification of Phytoremediation

The phytoremediation approach is classified based on the applicability, fundamental process, and type of pollutants into:

  i. Phytodegradation
 ii. Phytoextraction
iii. Phytostimulation or rhizodegradation
 iv. Phytovolatilization
  v. Rhizofiltration
 vi. Phytostabilization (Ismail 2012; Materac *et al.,* 2015; Sri Lakshmi *et al.,* 2017) [62 - 64].

Phytodegradation

Phytodegradation is called phytotransformation. Phytodegradation is a degradation process the enzymes are secreted from the plants which catalyze and enhance the chemical reaction in degrading the xenobiotics; this process can take place either inside or outside the plant (Alkio *et al.,* 2005; Burns *et al.,* 2013) [65, 66].

Around shoots and roots, microbes from plants are accumulated/colonized, subsequently, after accumulation, it can destroy the carbon substrates. Organic pollutants are besmirched into small molecules which will be bonded into the plant tissues and utilized for the growth of the plant.

Around the roots and shoot, these plant microbes are colonized and destroy the carbon substrates. Organic pollutants are converted into small molecules by degradation, later, these degraded small molecules are attached to the plant tissues and help for the growth of the plant (http://www.unep.or.jp/Ietc/Publications/ Freshwater/FMS2/2.asp; Materac *et al.,* 2015; Sri Lakshmi *et al.,* 2017) [63, 64, 67]. Plants such as Phreatophyte trees (willow, poplar, aspen, cottonwood), grasses (Bermuda, rye, fescue, sorghum), and legumes (alfalfa, clover, cowpeas) can be used for this process (Bharathiraja *et al.,* 2018) [19]. Organic contaminant degradation by the phytodegradation process (UNEP 2002) is shown in Fig. (**6**).

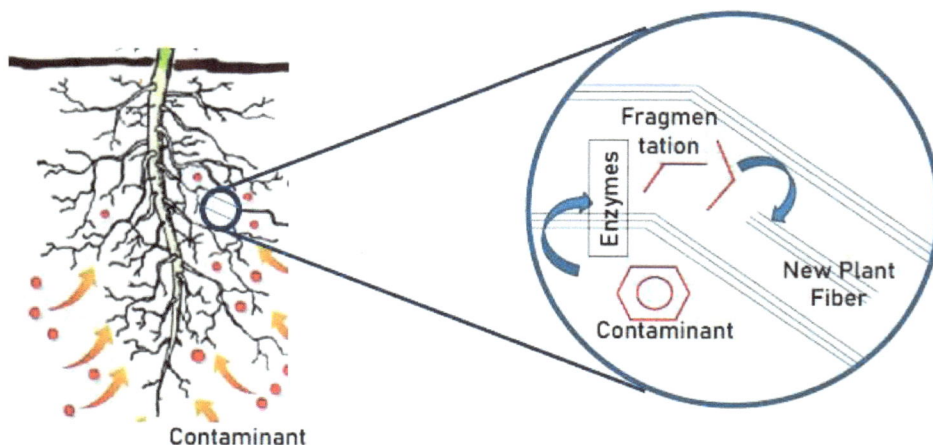

Fig. (6). Organic contaminant degradation by phytodegradation.

Phytoextraction

Mechanism - Absorption or dissolution in water and uptake or cation pumps accumulation or sequestration (Tahir *et al.,* 2016) [68].

Phytoextraction is also called phytoaccumulation. In this approach, plant roots uptake the toxic metal pollutants, transported and collected in the body part of the plant above the ground level. When the plant achieved its maximum growth, these accumulated metal components will be picked above the ground level, leading to the perpetual elimination of toxic metals from the particular site location. In this process metals such as copper, nickel and zinc can be eliminated (Raskin *et al.,* 1997; Ismail 2012; http://www.unep.or.jp/Ietc/Publications/Freshwater/FMS2/ 2.asp; 64-Materac *et al.,* 2015; Sri Lakshmi *et al.,* 2017) [63, 64, 67, 69]. Plants such as Indian mustard, sunflowers, barley, rapeseed plants, hops, serpentine plants, crucifers, nettles, and dandelions can be used for this process. Different metals such as Cd, As, Se, Pb, Zn, Cr, U with EDTA addition can be removed by this process (Bharathiraja *et al.,* 2018) [19]. Metal (Nickel) extraction by the phytoextraction process (http://www.unep.or.jp/Ietc/Publications/Freshwater /FMS2/2.asp) is shown in Fig. (**7**) [67].

Based on the nature, Phytoextraction process is classified as:

1. Chelate Assisted Phytoextraction

2. Continuous Phytoextraction

Phytoextraction depends on the following factors:

1. Quantity of metals in the soil

2. Plants absorbing capacity

3. Metal detoxification

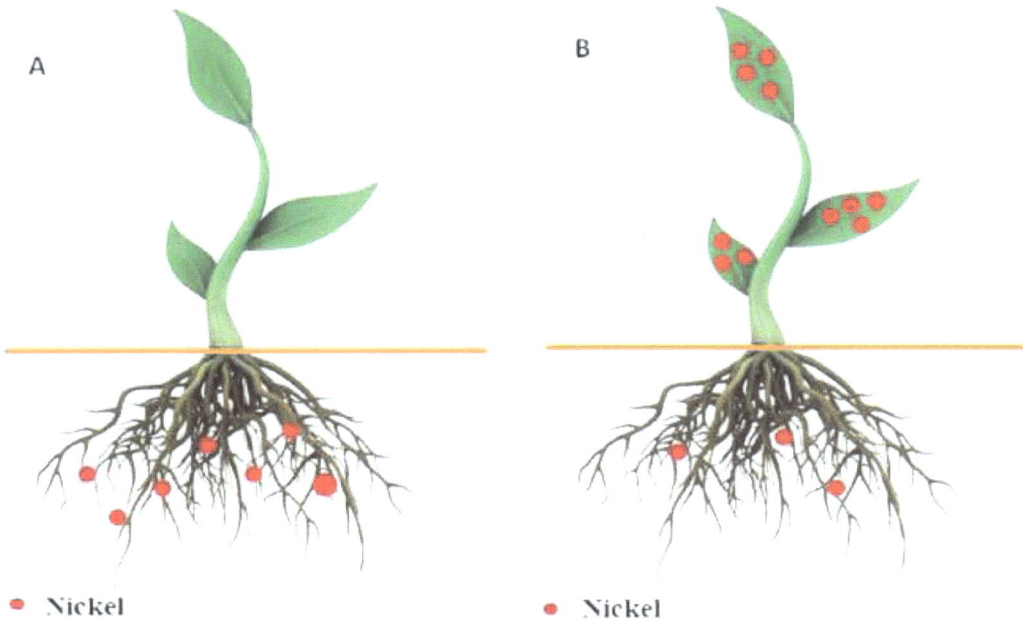

Fig. (7). (A). Before metal (Nickel) extraction by phytoextraction. **(B)**. After metal (Nickel) extraction by phytoextraction.

Phytostimulation

Mechanism – Root exudates or enzymes are secreted around the root zone followed by the degradation microbes of xenobiotics (Tahir *et al.,* 2016) [68].

Phytostimulation is also called rhizodegradation. Degradation of pollutants takes place through microorganisms that are present in the rhizosphere especially in plant roots surrounded by soil. This microorganism breaks the pollutants such as organic pollutants, fuels, solvents, and toxic compounds into simple and harmless compounds. These compounds are utilized as a nutrient source for their growth (Ismail 2012; http://www.unep.or.jp/Ietc/Publications/Freshwater/FMS2/2.asp; Materac *et al.,* 2015; Sri Lakshmi *et al.,* 2017) [63, 64, 67]. Plants such as phenolic releases like apple, orange, mulberry; fibrous root grasses like fescue, rye, and bermuda are used in this process. Organic Pollutants such as pesticides of aromatic and polynuclear aromatic hydrocarbons can be removed using this process (Bharathiraja *et al.,* 2018) [19].

Phytovolatilization

Mechanism - Alteration of pollutants/contaminants during vascular translocation from roots to leaves (Tahir *et al.,* 2016) [68].

In the phytovolatilization process, pollutants from the effluents are absorbed from the soil by the plant roots, transferring the absorbed pollutants to aerial parts of the plant and then to the leaves. Due to metabolic activities, these absorbed pollutants will be rehabilitated into volatile form (gaseous) and then transferred to the atmosphere. This process can take place in the tree growing stage and metals such as Se and Hg can be eliminated.

The main drawback of this method is the conversion of pollutants into volatile compounds that are then transferred to the atmosphere which is harmful (Ismail 2012; 68-http://www.unep.or.jp/Ietc/Publications/Freshwater/FMS2/2.asp; Materac *et al.,* 2015; Sri Lakshmi *et al.,* 2017) [63, 64, 67]. Plants such as Phreatophyte trees like willow, poplar, aspen, cottonwood; grasses like Bermuda, rye, fescue, sorghum; legumes like alfalfa, clover, cowpeas are used in this process. Pollutants such as chlorinated aliphatics, herbicides, nutrients, aromatics, and ammunition wastes can be removed by this process (Bharathiraja *et al.,* 2018) [19].

Phytofiltration

Mechanism - Filtration, sorption, or precipitation of pollutants surrounding the root zone.

Based on the plant parts, phytofiltration is named as:

• Plant roots as rhizofilteration
• Seedlings (blastofilteration) (Raskin *et al.,* 1997) [69].

The word "Rhizo" means "Root", roots of the plant absorb the metal pollutants in water rather than soil which is present next to the root zone.

This process is performed by either collecting the contaminated water from the waste site or hydroponically relocating into the contaminated water area, where contaminants are absorbed from the water through plant roots. After a certain period, plant root parts are saturated with the pollutants, consequently, whole plants or roots will be collected and disposed of. In this process, metals such as Cr, Zn and Pb can be removed.

Phytofiltration depends on the following factors:

1. Composition of Metals in the contaminants.

2. The capacity of plants metabolism (Ismail 2012; http://www.unep.or.jp/ Ietc/Publications/Freshwater/FMS2/2.asp; Materac *et al.,* 2015; Sri Lakshmi *et al.,* 2017) [63, 64, 67].

Phytostabilization

Mechanism -Sorption Precipitation or complexation in the rhizosphere.

Phytostabilization is also called phytorestoration. In this method, contaminants are restrained in the roots of the plant either physically in the root surface by sorption or chemically in the roots of the plant by chemical fixation.

Extreme root systems like absorption, adsorption, accumulation of pollutants/contaminants in the root parts, and tolerance to different pH are essential for this phytostabilization process (Ismail 2012; Materac *et al.,* 2015; Sri Lakshmi *et al.,* 2017; http://www.unep.or.jp/Ietc/Publications/Freshwater/FMS2/ 2.asp; Cunnigham and Berti 1993) [62 - 64, 67, 70]. Plants such as Phreatophyte trees; stabilization of soil erosion by grasses; sorb/bind contaminants dense root systems are needed. Metals such as Zn, Cd, As, Pb, Cr, Cu, U, Se, and hydrophobic organics like PCB, PAH, DDT, dieldrin can be removed by this process (Bharathiraja *et al.,* 2018) [19]. Different phytoremediation process (Pilon-Smith 2005) is shown in Fig. (**8**). [71].

Green Nanotechnology

Throughout the world, nanotechnology has become an innovative field in the 21st century. Traditionally nanoparticles are synthesized by physical and chemical methods. Recently nanoparticles were synthesized through a green approach (from different biological entities) that has been of great interest among the researchers because of avoiding toxic chemicals, cost-effective, single-step process, time-consuming, *etc.* Different biological entities like yeast, algae, bacteria, fungi, and plants are utilized in this green approach to synthesize metal nanoparticles and their application in effluent treatment, especially in textile dye treatment. Plants have been preferred as the best candidate among the different biological entities for the synthesis of metal NPs (Sangeeth and Kumaraguru 2013) [72].

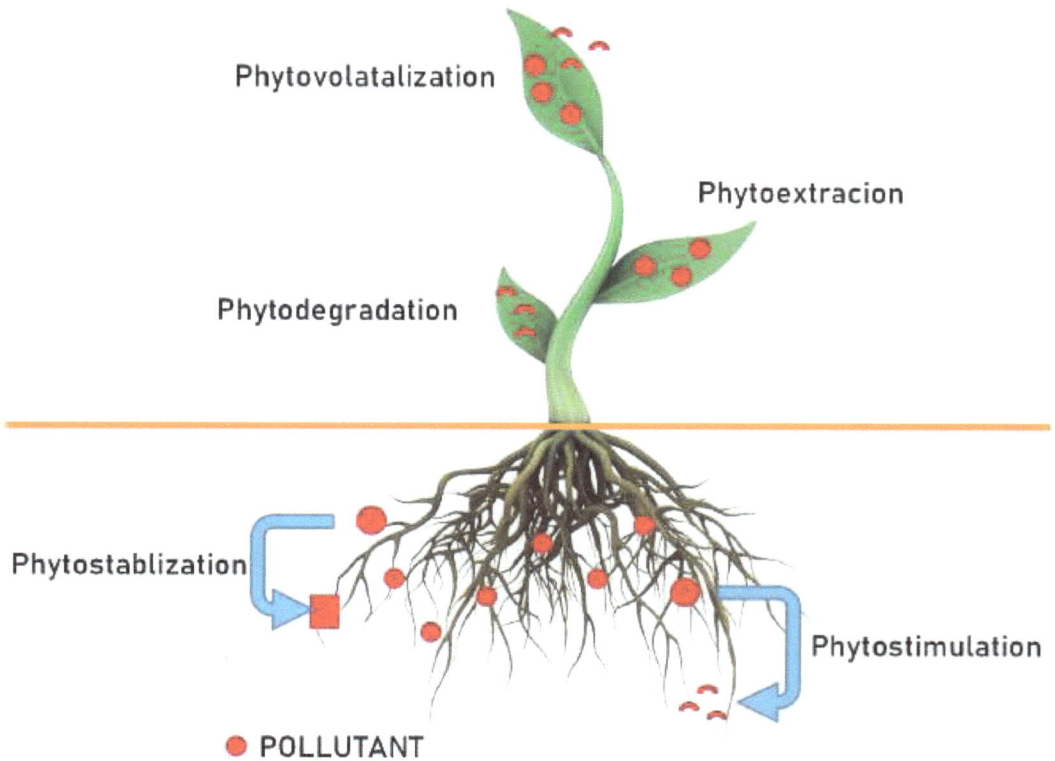

Fig. (8). Different phytoremediation process.

Different metal NPs such as Ag, Fe, Ni, Al, and Zn as zero-valent metal nanoparticles, TiO_2, Iron oxide, and ZnO as metal oxide nanoparticles, and other nanomaterials like carbon nanotubes and nanocomposites are widely utilized in the treatment of wastewater treatment. Green synthesized nanoparticles are considered to be non-toxic, cost-effective, and have long-term stability in nature. Various metal nanoparticles like silver, ZnO, Ion oxide, and TiO_2 have been synthesized by using different plants and investigated for their photo-degradation ability in different dyes and it has been reported that these metal nanoparticles have degraded the textile dyes. Among the different metal nanoparticles, the silver nanoparticle has been widely investigated for degrading the textile dyes (Singh *et al.,* 2016; Fairuzi *et al.,* 2018; Malhotra *et al.,* 2016; Bhakya *et al.,* 2015; Vanaja *et al.,* 2014; Jyoti and Singh 2016; Bonnia *et al.,* 2016; Zhou and Srinivasan 2015; Lu *et al.,* 2016) [73 - 81]. However, to develop a definite nanomaterial for the treatment of textile dye effluent, further studies are need to be performed to establish its efficiency, determine its potential toxicity and its impact on human health as well as the environmental needs.

CONCLUSION

The quality of the life of each living organism is decided based on its growth supporting systems for its survival in this polluted environment. Water has been treated as one of the prime resources for its survival. In simple without water, there is no life on earth. In the industrialization of urban areas, a huge volume of polluted water as effluent is discharged into the environment by different industries. Between the different industrial pollutants, textile dye industries effluent is considered as major pollutant/contaminant. The major objective in effluent treatment is the removal of pollutants/contaminants from the effluent, safeguarding the environment, and providing public health.

Several techniques like physical and chemical methods are utilized in effluent textile dye treatment but each one has its limitations. The main drawback of physical methods is that dye molecules are transferred to the other phase instead of eliminating the dye contaminant from the effluent; whereas, in chemical methods, a pretreatment with effective sludge production is needed. When compared with physical and chemical techniques, biological methods (phytoremediation) have numerous advantages like low cost, eco-friendly, simple implementation, and no secondary pollution.

With the advancement in science and technology in the last few decades, nanotechnology has made a new revolution almost in every field. Lately, plant-mediated synthesis (green approach) of nanoparticles has gained much interest among researchers in various fields including effluent treatment.

Researchers made numerous investigations towards the application of nanoparticles in dye effluent treatment and observed some promising results. Though various research investigations have been performed, still additional efforts need to be carried to launch a definite nanomaterial and establish its efficiency in treating textile dye effluent.

CONSENT FOR PUBLICATION

Not applicable.

CONFLICT OF INTEREST

The author declares no conflict of interest, financial or otherwise.

ACKNOWLEDGEMENT

Declared none.

REFERENCES

[1]   *Why is water so important to life on Earth.,* 2020.https://sciencing.com/about-6384365-watr-important-life-earth-.html

[2]   *Environmental health practitioner manual: A resource manual for environmental health practitioners working with aboriginal and torres strait islander communities.* http://www.health.gov.au/internet/ publications/publishing.nsf/Content/ohp-enhealth-manual-atsi-cnt-l~ohp-enhealth-manual-atsi-cnt-l-ch6~ohp-enhealth-manual-atsi-cnt-l-ch6.1

[3]   M.P. Shah, "Azo Dye Removal Technologies", *Austin J. Biotechnol. Bioeng.,* vol. 5, no. 1, p. 1090, 2018.

[4]   Z. Aksu, and G. Karabayır, "Comparison of biosorption properties of different kinds of fungi for the removal of Gryfalan Black RL metal-complex dye", *Bioresour. Technol.,* vol. 99, no. 16, pp. 7730-7741, 2008.
[http://dx.doi.org/10.1016/j.biortech.2008.01.056] [PMID: 18325761]

[5]   T. Robinson, G. McMullan, R. Marchant, and P. Nigam, "Remediation of dyes in textile effluent: a critical review on current treatment technologies with a proposed alternative", *Bioresour. Technol.,* vol. 77, no. 3, pp. 247-255, 2001.
[http://dx.doi.org/10.1016/S0960-8524(00)00080-8] [PMID: 11272011]

[6]   Z. Aksu, and S. Tezer, "Biosorption of reactive dyes on the green alga Chlorella vulgaris", *Process Biochem.,* vol. 40, no. 3-4, pp. 1347-1361, 2005.
[http://dx.doi.org/10.1016/j.procbio.2004.06.007]

[7]   H. Alhassani, M. Rauf, and S. Ashraf, "Efficient microbial degradation of Toluidine Blue dye by Brevibacillus sp", *Dyes Pigments,* vol. 75, no. 2, pp. 395-400, 2007.
[http://dx.doi.org/10.1016/j.dyepig.2006.06.019]

[8]   R. Palanivelan, S. Rajakumar, P. Jayanthi, and P.M. Ayyasamy, "Potential process implicated in bioremediation of textile effluents: A review", *Adv. Appl. Sci. Res.,* vol. 4, no. 1, pp. 131-145, 2013.

[9]   R. Shivangi, *"Microbial Degradation and Treatment Studies on Textile Waste Water of Bagru Region (Rajasthan). Doctoral Thesis"*, Jayoti Vidyapeeth Women's University., Jaipur (Rajasthan), India, 2012.

[10]   C.R. Holkar, A.J. Jadhav, D.V. Pinjari, N.M. Mahamuni, and A.B. Pandit, "A critical review on textile wastewater treatments: Possible approaches", *J. Environ. Manage.,* vol. 182, pp. 351-366, 2016.
[http://dx.doi.org/10.1016/j.jenvman.2016.07.090] [PMID: 27497312]

[11]   B. Rani, R. Maheshwari, R.K. Yadav, D. Pareek, and A. Sharma, *Resolution to provide safe drinking water for sustainability of future perspectives,* vol. 1, pp. 50-54, 2013.

[12]   J.P. Jadhav, S.S. Phugar, R.S. Dhanve, and S.B. Jadhav, "Rapid biodegradation and decolorization of Direct Orange 39 (Orange TGLL) by an isolated bacterium Pseudomonas aeruginosa strain BCH", *Biodegrdation.,* vol. 21, no. 3, pp. 453-463, 2010.
[http://dx.doi.org/10.1007/s10532-009-9315-6]

[13]   R.V. Khandare, and S.P. Govindwar, *Phytoremediation of textile dyes and effluents: Current scenario and future prospects* vol. 33. Biotech. Adv, 2015, pp. 1697-1714.

[14]   K.S. Kimmatkar, A.V. Purohit, and A.J. Sanyal, "Phytoremediation Techniques and Species for Combating Contaminants of Textile Effluents – An Overview", *Int. J. Sci. Res.,* vol. 6, no. 3, pp. 2174-2181, 2017.

[15]   A.M. Talarposhti, T. Donnelly, and G.K. Anderson, "Colour removal from a simulated dye wastewater using a two-phase Anaerobic packed bed reactor", *Water Res.*, vol. 35, no. 2, pp. 425-432, 2001.
[http://dx.doi.org/10.1016/S0043-1354(00)00280-3] [PMID: 11228995]

[16]   V.K. Gupta, I. Ali, Suhas, and D. Mohan, "Equilibrium uptake and sorption dynamics for the removal of a basic dye (basic red) using low-cost adsorbents", *J. Colloid Interface Sci.*, vol. 265, no. 2, pp. 257-264, 2003.
[http://dx.doi.org/10.1016/S0021-9797(03)00467-3] [PMID: 12962659]

[17]   M.A.A. Hassan, T.P. Li, and Z.Z. Noor, "Coagulation and flocculation treatment of wastewater in textile industry using chitosan", *J. Chem. Nat. Resour. Eng.*, vol. 4, no. 1, pp. 43-53, 2009.

[18]   H.L. Chen, and L.D. Burns, "Environmental analysis of textile products", *Cloth. Text. Res. J.*, vol. 24, no. 3, pp. 248-261, 2006.
[http://dx.doi.org/10.1177/0887302X06293065]

[19]   B. Bharathiraja, J. Jayamuthunagai, R. Praveenkumar, and J. Iyyappan, "Phytoremediation Techniques for the Removal of Dye in Wastewater", In: *Bioremediation: Applications for Environmental Protection and Management, Energy, Environment, and Sustainability.* Springer Nature Singapore Pvt Ltd., 2018, pp. 243-252.

[20]   H.M. Pinheiro, E. Touraud, and O. Thomas, "Aromatic amines from azo dye reduction: status review with emphasis on direct UV spectrophotometric detection in textile industry wastewaters", *Dyes Pigments*, vol. 61, no. 2, pp. 121-139, 2004.
[http://dx.doi.org/10.1016/j.dyepig.2003.10.009]

[21]   Y.C. Chung, and C.Y. Chen, "Degradation of azo dye reactive violet 5 by TiO2 photocatalysis", *Environ. Chem. Lett.*, vol. 7, no. 4, pp. 347-352, 2009.
[http://dx.doi.org/10.1007/s10311-008-0178-6]

[22]   H. Ali, "Biodegradation of synthetic dyes-a review", *Water Air Soil Pollut.*, vol. 213, no. 1-4, pp. 251-273, 2010.
[http://dx.doi.org/10.1007/s11270-010-0382-4]

[23]   P. Mahajan, J. Kaushal, A. Upmanyu, and J. Bhatti, "Assessment of Phytoremediation Potential of Chara vulgaris to Treat Toxic Pollutants of Textile Effluent", *J. Toxicol.*, vol. 2019, pp. 1-11, 2019.
[http://dx.doi.org/10.1155/2019/8351272] [PMID: 30853979]

[24]   Metcalf and Eddy, *Wastewater Engineering: Treatment, Disposal and Reuse.* 3rd ed. McGraw Hill: New York, 1991.

[25]   S. Lakshmana Prabu, T.N.K. Suriyaprakash, K. Ruckmani, and R. Thirumurugan, "Effective Waste Water Treatment and its Management", In: *Toxicity and Waste Management Using Bioremediation.* IGI Global, 2016, pp. 312-334.
[http://dx.doi.org/10.4018/978-1-4666-9734-8.ch016]

[26]   A. Ahmad, S. Siti, C. Sing, K. Asma, A.W. Waseem, and K. Rajeev, "Recent advances in new generation dye removal technologies: novel search of approaches to reprocess waste water", *RSC Advances*, vol. 39, pp. 1-58, 2015.

[27]   *What is membrane separation?*.https://www.asahi-kasei.co.jp/membrane/microza/en/kiso/kiso_1.html

[28]   *Membrane materials.* http://www.separationprocesses.com/Membrane/MT_Chp03.htm

[29]   D.R. Mohan, J. Pramila, E. Kavitha, D. Nithya, G. Hema Mala, and M. Tarun, "Remediation of textile effluents by membrane based treatment techniques: A state of the art review", *J. Chem. Pharma. Sci.*, vol. 4, pp. 296-299, 2014.

[30]   *Principles of flocculation dispersion selective flocculation.*, 2020.http://www.columbia.edu/~ps24/PDFs/Principles%20of%20Flocculation%20Dispersion%20Selective %20Flocculation.pdf

[31]   M.A. Sabur, A.A. Khan, and S. Safiullah, "Treatment of textile wastewater by coagulation precipitation method", *Journal of Scientific Research*, vol. 4, no. 3, pp. 623-633, 2012.

[http://dx.doi.org/10.3329/jsr.v4i3.10777]

[32]   L.N. Ukiwe, S.I. Ibeneme, C.E. Duru, B.N. Okolue, G.O. Onyedika, and C.A. Nweze, "Chemical and electro-coagulation techniques in coagulation floccculation in water and wastewater treatment- A review", *J. Adv. Chem.*, vol. 9, no. 3, pp. 1988-1999, 2014.

[33]   F. Mazille, and D. Spuhler, https://sswm.info/sswm-university-course/module-6-disaster-situa-ions-planning-and-preparedness/further-resources-0/coagulation-flocculation

[34]   Base Exchange (Ion Exchange) Water Softening Process, https://textilelearner.blogspot.com /2012/05/base-exchange-ion-exchange-water.html

[35]   V. Parameswaran, E.R. Nagarajan, and A. Murugan, "Treatment of textile effluents by ion-exchange polymeric materials", *Int. J. Chemtech Res.*, vol. 6, no. 6, pp. 3332-3335, 2014.

[36]   M. Wawrzkiewicz, and Z. Hubicki, "Anion Exchange Resins as Effective Sorbents for Removal of Acid, Reactive, and Direct Dyes from Textile Wastewaters", In: *Ion Exchange - Studies and Applications, IntechOpen.* Crotia, 2015, pp. 37-72.
[http://dx.doi.org/10.5772/60952]

[37]   R. Pesoutova, P. Hlavinek, and J. Matysikova, "Use of advanced oxidation processes for textile wastewater treatment – a review", In: *Food Environ. Safety – J. Faculty Food Eng.* vol. 10. , 2011, no. 3, pp. 59-65.

[38]   D. Huang, C. Hu, G. Zeng, M. Cheng, P. Xu, X. Gong, R. Wang, and W. Xue, "Combination of Fenton processes and biotreatment for wastewater treatment and soil remediation", *Sci. Total Environ.*, vol. 574, pp. 1599-1610, 2017.
[http://dx.doi.org/10.1016/j.scitotenv.2016.08.199] [PMID: 27608610]

[39]   M. Pera-Titus, V. García-Molina, M.A. Baños, J. Giménez, and S. Esplugas, "Degradation of chlorophenols by means of advanced oxidation processes: a general review", *Appl. Catal. B*, vol. 47, no. 4, pp. 219-256, 2004.
[http://dx.doi.org/10.1016/j.apcatb.2003.09.010]

[40]   K.Y. Foo, and B.H. Hameed, "An overview of dye removal *via* activated carbon adsorption process", *Desalination Water Treat.*, vol. 19, no. 1-3, pp. 255-274, 2010.
[http://dx.doi.org/10.5004/dwt.2010.1214]

[41]   A.K. Popuri, and P. Guttikonda, "Treatment of textile dyeing industry effluent using activated carbon", *Int. J. Chem. Sci.*, vol. 13, no. 3, pp. 1430-1436, 2015.

[42]   H. Patel, "Charcoal as an adsorbent for textile wastewater treatment", *Sep. Sci. Technol.*, vol. 53, no. 17, pp. 2797-2812, 2018.
[http://dx.doi.org/10.1080/01496395.2018.1473880]

[43]   B. Shriram, and S. Kanmani, "Ozonation of textile dyeing wastewater -", *RE:view*, vol. 15, no. 3, pp. 46-50, 2014.

[44]   M.A.R. Bhuiyan, M.M. Rahman, A. Shaid, and M.A. Khan, "Application of gamma irradiated textile wastewater for the pretreatment of cotton fabric", *Environment and Ecology Research*, vol. 2, no. 3, pp. 149-152, 2014.
[http://dx.doi.org/10.13189/eer.2014.020304]

[45]   F. Parvin, Z. Ferdaus, S.M. Tareq, T.R. Choudhury, J.M.M. Islam, and M.A. Khan, "Effect of gamma-irradiated textile effluent on plant growth", *Int. J. Recycl. Org. Waste Agric.*, vol. 4, no. 1, pp. 23-30, 2015.
[http://dx.doi.org/10.1007/s40093-014-0081-z]

[46]   Y. Feng, L. Yang, J. Liu, and B.E. Logan, "Electrochemical technologies for wastewater treatment and resource reclamation", *Environ. Sci. Water Res. Technol.*, vol. 2, no. 5, pp. 800-831, 2016.
[http://dx.doi.org/10.1039/C5EW00289C]

[47]   J. Radjenovic, and D.L. Sedlak, "Challenges and Opportunities for electrochemical processes as next-

generation technologies for the treatment of contaminated water", *Environ. Sci. Technol.,* vol. 49, no. 19, pp. 11292-11302, 2015.
[http://dx.doi.org/10.1021/acs.est.5b02414] [PMID: 26370517]

[48]   C.A. Martínez-Huitle, and L.S. Andrade, "Electrocatalysis in wastewater treatment: recent mechanism advances", *Quim. Nova,* vol. 34, no. 5, pp. 850-858, 2011.
[http://dx.doi.org/10.1590/S0100-40422011000500021]

[49]   P. Cañizares, M. Carmona, J. Lobato, F. Martínez, and M.A. Rodrigo, "Electro dissolution of aluminum electrodes in electrocoagulation processes", *Ind. Eng. Chem. Res.,* vol. 44, no. 12, pp. 4178-4185, 2005.
[http://dx.doi.org/10.1021/ie048858a]

[50]   A. Arulmurugan, K. Chithra, R. Thilakavathi, and N. Balasubramanian, "Degradation of textile effluent by electro coagulation technique", *Bull. Electrochem.,* vol. 23, pp. 247-252, 2007.

[51]   E. Butler, Y.T. Hung, R.Y.L. Yeh, and M. Suleiman Al Ahmad, "Electrocoagulation in wastewater treatment", *Water,* vol. 3, no. 2, pp. 495-525, 2011.
[http://dx.doi.org/10.3390/w3020495]

[52]   L.N. Ukiwe, S.I. Ibeneme, C.E. Duru, B.N. Okolue, G.O. Onyedika, and C.A. Nweze, "Chemical and electro-coagulation techniques in coagulation-floccculation in water and wastewater treatment- A review", *J. Adv. Chem.,* vol. 9, no. 3, pp. 1988-1999, 2014.

[53]   A.S. Naje, S. Chelliapan, Z. Zakaria, and S.A. Abbas, "Treatment performance of textile wastewater using electrocoagulation (EC) process under combined electrical connection of electrodes", *Int. J. Electrochem. Sci.,* vol. 10, pp. 5924-5941, 2015.

[54]   S.H. Abbas, and W.H. Ali, "Electrocoagulation technique used to treat wastewater: A review", *Am. J. Eng. Res.,* vol. 7, no. 10, pp. 74-88, 2018.

[55]   P.S. Kore, V.C. Mugale, N.S. Kulal, S.P. Thaware, A.M. Vanjuari, and K.M. Mane, "Textile waste water treatment by using phytoremediation", *Int. J. Engg. Trends Tech.,* vol. 45, no. 8, pp. 412-415, 2017.
[http://dx.doi.org/10.14445/22315381/IJETT-V45P276]

[56]   K.H.D. Tang, N.M. Darwish, A.M. Alkahtani, M.R.A. Gawwad, P. Karacsony, "Biological Removal of Dyes from Wastewater: A Review of Its Efficiency and Advances", *Trop. Aquatic Soil Pollut.,* vol.2, no.1, pp. 59-75, 2022.
[http://dx.doi.org/10.1016/S0167-7799(00)01514-6] [PMID: 11146098]

[57]   T. Macek, M. Macková, and J. Káš, "Exploitation of plants for the removal of organics in environmental remediation", *Biotechnol. Adv.,* vol. 18, no. 1, pp. 23-34, 2000.
[http://dx.doi.org/10.1016/S0734-9750(99)00034-8] [PMID: 14538117]

[58]   S. Susarla, V.F. Medina, and S.C. McCutcheon, "Phytoremediation: An ecological solution to organic chemical contamination", *Ecol. Eng.,* vol. 18, no. 5, pp. 647-658, 2002.
[http://dx.doi.org/10.1016/S0925-8574(02)00026-5]

[59]   H. Xia, L. Wu, and Q. Tao, "[A review on phytoremediation of organic contaminants]", *Ying Yong Sheng Tai Xue Bao,* vol. 14, no. 3, pp. 457-460, 2003.
[PMID: 12836561]

[60]   M. Vasanthy, M. Santhiya, V. Swabna, and A. Geetha, "Phytodegradation of textile dyes by Water Hyacinth (Eichhornia Crassipes) from aqueous dye solutions", *Int. J. Environ. Sci.,* vol. 1, no. 7, pp. 1702-1717, 2011.

[61]   J.L. Schnoor, L.A. Licht, S.C. McCUTCHEON, N.L. Wolfe, and L.H. Carreira, "Phytoremediation of organic and nutrient contaminants", *Environ. Sci. Technol.,* vol. 29, no. 7, pp. 318A-323A, 1995.
[http://dx.doi.org/10.1021/es00007a747] [PMID: 22667744]

[62]   S. Ismail, ""Phytoremediation: a green technology", Ira", *J. Plant Physiol.,* vol. 3, no. 1, pp. 567-576, 2012.

[63]   M. Materac, A. Wyrwicka, and E. Sobiecka, "Phytoremediation techniques in wastewater treatment", *Environ. Biotechnol.,* vol. 11, no. 1, pp. 10-13, 2015.
[http://dx.doi.org/10.14799/ebms249]

[64]   K.S. Lakshmi, V.H. Sailaja, and M.A. Reddy, "Phytoremediation - A promising technique in waste water treatment", *International Journal of Scientific Research and Management,* vol. 5, no. 6, pp. 5480-5489, 2017.
[http://dx.doi.org/10.18535/ijsrm/v5i6.20]

[65]   M. Alkio, T.M. Tabuchi, X. Wang, and A. Colón-Carmona, "Stress responses to polycyclic aromatic hydrocarbons in Arabidopsis include growth inhibition and hypersensitive response-like symptoms", *J. Exp. Bot.,* vol. 56, no. 421, pp. 2983-2994, 2005.
[http://dx.doi.org/10.1093/jxb/eri295] [PMID: 16207747]

[66]   R.G. Burns, J.L. DeForest, J. Marxsen, R.L. Sinsabaugh, M.E. Stromberger, M.D. Wallenstein, M.N. Weintraub, and A. Zoppini, "Soil enzymes in a changing environment: Current knowledge and future directions", *Soil Biol. Biochem.,* vol. 58, pp. 216-234, 2013.
[http://dx.doi.org/10.1016/j.soilbio.2012.11.009]

[67]   *Phytoremediation: An environmentally sound technology for pollution prevention, control and redmediation.* http://www.unep.or.jp/Ietc/Publications/Freshwater/FMS2/2.asp

[68]   U. Tahir, A. Yasmin, and U.H. Khan, "Phytoremediation: Potential flora for synthetic dyestuff metabolism", *J. King Saud Univ. Sci.,* vol. 28, no. 2, pp. 119-130, 2016.
[http://dx.doi.org/10.1016/j.jksus.2015.05.009]

[69]   I. Raskin, R.D. Smith, and D.E. Salt, "Phytoremediation of metals: using plants to remove pollutants from the environment", *Curr. Opin. Biotechnol.,* vol. 8, no. 2, pp. 221-226, 1997.
[http://dx.doi.org/10.1016/S0958-1669(97)80106-1] [PMID: 9079727]

[70]   S.D. Cunningham, and W.R. Berti, "Remediation of contaminated soils with green plants: An overview", *In Vitro Cell. Dev. Biol. Plant,* vol. 29, no. 4, pp. 207-212, 1993.
[http://dx.doi.org/10.1007/BF02632036]

[71]   E. Pilon-Smits, "Phytoremediation", *Annu. Rev. Plant Biol.,* vol. 56, no. 1, pp. 15-39, 2005.
[http://dx.doi.org/10.1146/annurev.arplant.56.032604.144214] [PMID: 15862088]

[72]   N. Sangeeth, and A.K. Kumaraguru, "Extracellular synthesis of zinc oxide nanoparticle using seaweeds of gulf of Mannar, India", *Nanobiotechnol.,* vol. 11, pp. 1-11, 2013.

[73]   T. Singh, K. Jyoti, A. Patnaik, R. Chauhan, and N. Kumar, *Application of silver nanoparticles synthesized from Raphanus sativus for catalytic degradation of organic dyes,* p. 05003, 2016.

[74]   A.A. Fairuzi, N.N. Bonnia, R.M. Akhir, M.A. Abrani, and H.M. Akil, "Degradation of methylene blue using silver nanoparticles synthesized from imperata cylindrica aqueous extract", *IOP Conf. Ser. Earth Environ. Sci.,* vol. 105, p. 012018, 2018.
[http://dx.doi.org/10.1088/1755-1315/105/1/012018]

[75]   P. Malhotra, R. Kathal, and A. Puri, "Iron nanoparticles catalyzed degradation of organic dyes in water for environmental remediation", *J. Basic Appl. Eng. Res.,* vol. 3, no. 1, pp. 41-43, 2016.

[76]   S. Bhakya, S. Muthukrishnan, M. Sukumaran, M. Muthukumar, T. Senthil Kumar, and M.V. Rao, "Catalytic degradation of organic dyes using synthesized silver nanoparticles: A Green approach", *J. Bioremediat. Biodegrad.,* vol. 6, p. 313, 2015.

[77]   M. Vanaja, K. Paulkumar, M. Baburaja, S. Rajeshkumar, G. Gnanajobitha, C. Malarkodi, M. Sivakavinesan, and G. Annadurai, "Degradation of methylene blue using biologically synthesized silver nanoparticles", *Bioinorg. Chem. Appl.,* vol. 2014, pp. 1-8, 2014.
[http://dx.doi.org/10.1155/2014/742346] [PMID: 24772055]

[78]   K. Jyoti, and A. Singh, "Green synthesis of nanostructured silver particles and their catalytic application in dye degradation", *J. Genet. Eng. Biotechnol.,* vol. 14, no. 2, pp. 311-317, 2016.

[http://dx.doi.org/10.1016/j.jgeb.2016.09.005] [PMID: 30647629]

[79]   N.N. Bonnia, M.S. Kamaruddin, M.H. Nawawi, S. Ratim, H.N. Azlina, and E.S. Ali, "Green biosynthesis of silver nanoparticles using Polygonum Hydropiper and study its catalytic degradation of methylene blue", *Procedia Chem.,* vol. 19, pp. 594-602, 2016.
[http://dx.doi.org/10.1016/j.proche.2016.03.058]

[80]   R. Zhou, and M.P. Srinivasan, "Photocatalysis in a packed bed: Degradation of organic dyes by immobilized silver nanoparticles", *J. Environ. Chem. Eng.,* vol. 3, no. 2, pp. 609-616, 2015.
[http://dx.doi.org/10.1016/j.jece.2015.02.004]

[81]   H. Lu, J. Wang, M. Stoller, T. Wang, Y. Bao, and H. Hao, "An overview of nanomaterials for water and wastewater treatment", *Adv. Mater. Sci. Eng.,* vol. 2016, pp. 1-10, 2016.
[http://dx.doi.org/10.1155/2016/4964828]

CHAPTER 5

Composting of Fruit Wastes - An Efficient and Alternative Option for Solid Waste Management

Anbarasi Karunanithi[1,*] and Dhanaraja Dhanapal[2]

[1] *University College of Engineering, BIT Campus, Anna University, Trichy - 620 024, India*

[2] *Paavai Engineering College, Pachal, Namakkal - 637018, India*

Abstract: Municipal solid waste generation is exponentially increasing every year. Managing these solid wastes is highly complicated due to the generation of a plethora of waste. Since collecting and disposing of wastes in dumping sites cause severe environmental impacts, an alternative option is the need of the hour. Thus the technique used must be efficient and less in cost for agricultural applications. Composting is one such process where the decomposition and recycling of organic material into a humus-rich soil take place naturally known as compost. Fruit waste is rich in moisture content, and thus possesses a unique property as a raw compost agent. The present study focuses on composting of fruit wastes for reducing the amount of solid waste being collected and dumped. If composting of fruit waste is carried out in backyards, then the amount of solid waste entering the dumping sites can be reduced substantially.

Keywords: Composting, Humus-rich soil, Municipal solid waste.

INTRODUCTION

Solid waste disposal is one of the primary environmental problems faced by the country presently since it destroys both the environment and the ecological cycle [1]. The contribution of solid wastes is estimated to be about 40% of organic wastes which can be reused and converted into environmentally compatible products [2]. When these organic wastes are left unattended, two adverse effects can occur. Firstly, the unbearable smell it causes due to natural decomposition and secondly, it gives rise to severe health problems by feeding insects and pests. Hence the recycling or reuse of organic waste is necessary.

Food waste is one of the major contributors to solid waste, and when it is not composted, it goes directly to a landfill. This organic matter may react with other

---

* Corresponding author Anbarasi Karunanithi: University college of Engineering, BIT Campus, Anna University, Trichy - 620 024, India; E-mail: kanbarasi@aubit.edu.in

G. Venkatesan, S. Lakshmana Prabu and M. Rengasamy (Eds.)

materials in thelandfill and create toxic leachate [3]. Food waste has the capacity to stop the earth's natural cycle of decomposition when placed in an airtight landfill. Composting can be done by separating, segregating and decomposing by biological means [3, 4]. Composting is considered to be the most economical and sustainable option for organic waste management [5] since it provides a way in which solid wastes, water quality, and agricultural concerns can be joined. This compost not only increases the water and nutrient holding capacity, thereby improving the soil quality but also reduces waste going to landfills, improves plant productivity, and reduces water runoff and soil erosion [6].

Food waste is rich in moisture content and thus possesses a unique property as a raw compost agent. Hence it is important to mix fresh food waste with a bulking agent that will absorb excess moisture as well as add structure to the mixture [7]. Many factors play an active role in the composting process. Some of the factors are C/N ratio, moisture content, temperature, oxygen content, particle size, porosity and bulk density [8]. The loss of nutrients and the time required for composting can be minimized by optimizing these operating parameters [9].

At present, an environmentally friendly technology is most important for utilizing organic waste. Composting is one such important and efficient method of solid waste management [10]. The compost not only reduces the amount of waste but it can also be used for improving soil fertility; hence reducing the amount of fertilizer to be used in the land [2]. The present study aimed to compost fruit waste to reduce the amount of solid waste being collected and dumped.

COMPOSTING METHODS

There are various methods of composting. Some of them are given below:

Bin Composting is typically used for small amounts of food waste. They require little labor, use wire mesh or wooden frames for better air circulation, and are inexpensive. Three chamber bins that can handle significant quantities of materials are used for faster compost production utilizing varying stages of decomposition. This bin composting process also allows staged composting, with one section used for storing compostable materials, one section for active composting, and one section for finished compost [1].

Vermi Composting is the most commonly used composting process that uses worms to consume food waste. This type is usually done in containers or bins as shown in Fig. (**1**). As an environmental education tool, many educational institutions use this method for high quality compost. One of the main disadvantages is that the investment in worm stocking may be high depending on the size of the operation. The investment in worm stocking may be high depending on the size of the operation. Also, if too much waste is added, anaerobic conditions may occur [2].

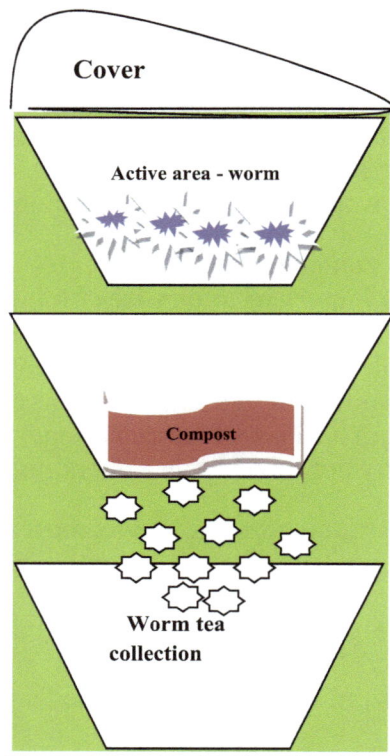

Fig. (1). Vermicomposting.

Aerated Static Piles Method- In this method, there are three layers, namely, plenum, active and biofiltration layer. The air is introduced to the stacked pile using perforated pipes and blowers as shown in Fig. (**2**). It is then slowed and diffused in the plenum layer and the active layer promotes aerobic composting. The final biofilteration layer traps heat and prevents the escape of odour. Since this method is weather sensitive, it requires no labor and due to imperfect mixing, it can have unreliable pathogen reduction [2].

Windrows Method is used for larger volumes and requires huge space. This method produces a uniform product and can be remotely located. The cost equipment used for this process is expensive. Fig. (**4**) shows that in this process, long, narrow piles are turned when required based on temperature and oxygen requirements. The only disadvantage of this process is it can have odor problems, and when it is exposed to rainfall, it can create leachate problems [10].

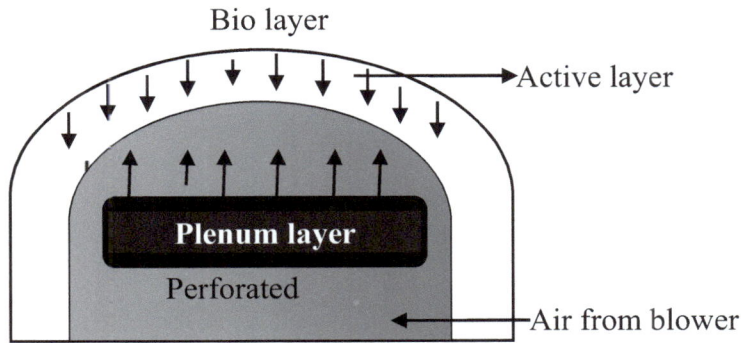

Fig. (2). Aerated static piles method.

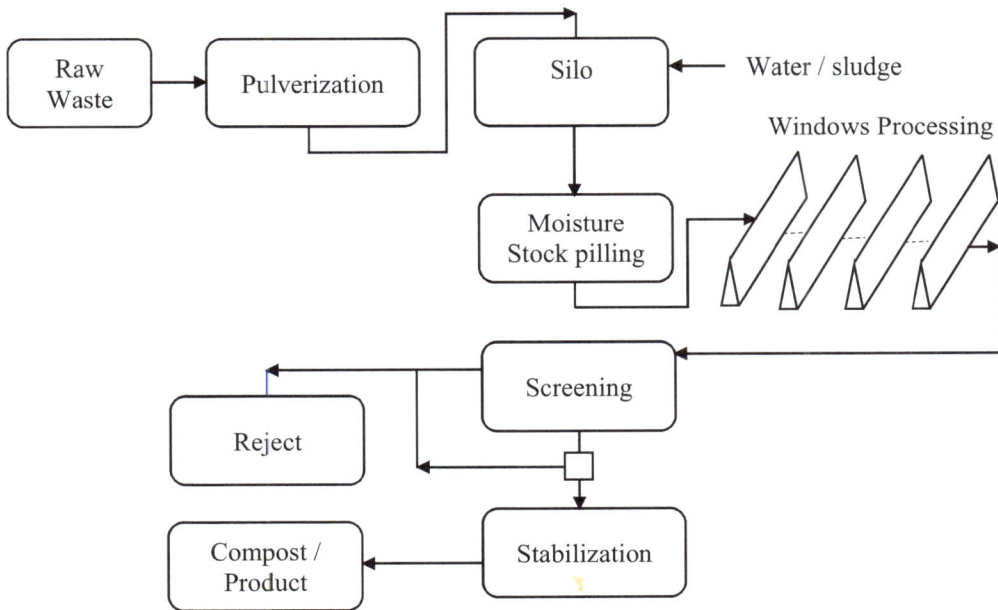

Fig. (3). Windrows method.

Passive Composting is a natural decomposition process that involves stacking the materials and leaving them to compost over a long period of time. It is one of the most common methods used nowadays for decomposition. This method is a simple, low cost method where the process is very slow and may result in objectionable odors. Since this process is natural, it depends on nature, where cool air and oxygen are drawn into the pile and the warm air is released. This process is commonly referred to as the chimney effect [1].

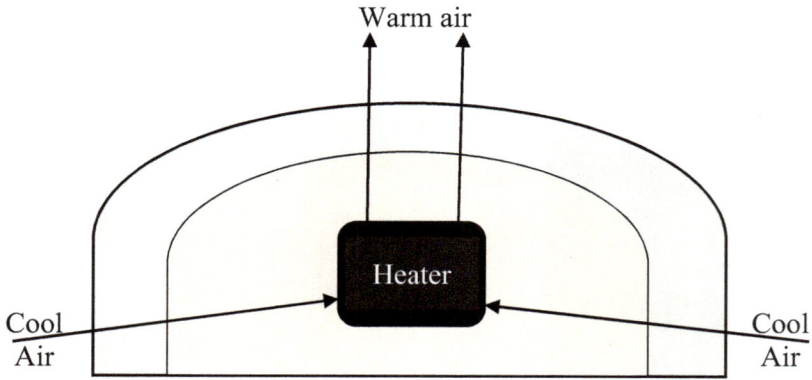

Fig. (4). Passive composting process.

In-Vessel Systems can be used both in urban and public areas. They are simple to use with minimal labor. These systems are usually perforated barrels or drums or specially manufactured containers that are easy to turn. The initial investment can be high and handling volumes are typically low. These in vessel systems are not weather sensitive [10]. A typical in – vessel system is shown in Fig. (5).

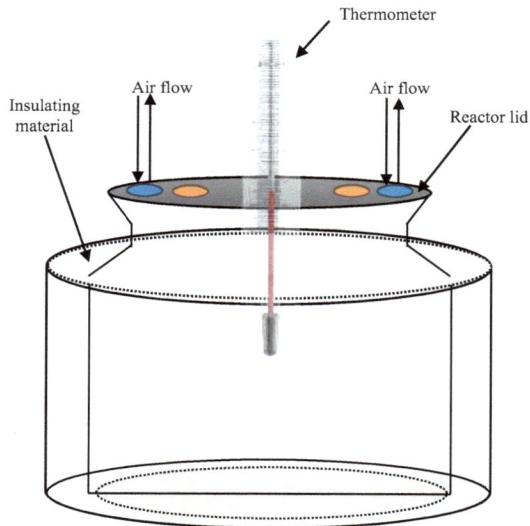

Fig. (5). In-vessel systems.

INFLUENCE OF PROCESS PARAMETERS IN THE COMPOSTING PROCESS

Influence of Temperature: of the compost plays a vital role in the decomposition process. Generally high temperatures destroy weed seeds and pathogens in the compost. Hence temperature should neither be low nor high as low outside temperature slows down the process, while warmer conditions speed up the process. To begin the composting process, mesophilic bacteria function between 50 and 113 degrees F whereas thermophilic bacteria take over and thrive between 113 to 158 degrees F. Sometimes composting manures can reach temperatures as high as 200 degrees F. However, temperatures above 158 degrees F may char the compost or create conditions suitable for combustion spontaneously [8, 12, 13].

Influence of Oxygen Content: Aeration is an essential process for optimum microorganism populations to effectively break down the composting material. Aeration can be done by blowers, fans, aeration tubes, aeration holes, or raising the compost off the ground [7].

Influence of PH: For an efficient decomposition, a fairly neutral pH will ensure high levels of microorganisms and a proper carbon – nitrogen ratio (C: N) should create optimum pH levels. Normally PH ranges from 6.0 to 7.8 are considered as high quality compost [11].

Influence of Moisture Content: For a microorganism to break down the compost, an optimum moisture content of 60% is required. If it is above 70%, the process will slow down and can create a foul smell due to anaerobic conditions. Moisture below 50% also slows down the decomposition process. The ideal moisture content of fresh food waste is said to be from 80 to 90%, sawdust is 25%, and yard waste is 70%. Compost should not be too wet or too dry. If the clump is too wet, then it drips water, so more aeration is required. If the compost falls through your fingers, it is too dry, so more food waste or water can be added [8].

Influence of Particle Size: is one of the process parameters that can affect the rate of decomposition of compost. When the particles are small, the compost receives more aeration. As a result, microorganisms can break down smaller pieces faster. Smaller particle sizes of composted materials can be achieved by various methods like shredding, chipping and cutting before they enter the compost pile [12].

Influence of Carbon to Nitrogen Ratio (C/N): In order to process an organic material to compost, bacteria present in compost need a proper nutrient mix. The optimum C/N ratio to begin composting is 30:1. When the C/N ratio increases, decomposition is slowed. Whereas foul smell and loss of nitrogen occur when the ratio decreases. The ideal ratio for food waste is 15:1, fruit waste 35:1, leaves 60:1, bark 100:1, and sawdust 500:1, respectively [11].

Table 1. Acceptable and ideal limits of process parameters to be considered in composting.

Process Parameters			
S. No.	**-**	**Acceptable Limit**	**Ideal Limit**
1	Temperature	40 – 60 º C	50 – 60 °C
2	Oxygen content (Aeration)	≥ 6%	≥ 12%
3	PH level	5 – 9	6 – 8
4	Particle size	12 – 25 mm	≤ 25 mm
5	Moisture content (by weight)	40 – 60%	45 – 60%
6	Carbon to nitrogen ratio (C/N)	20:1 – 40:1	25:1 – 35:1
7	Porosity	40 – 60%	40 – 60%
8	Bulk density	600 kg/m^3	600 kg/m^3

CONCLUSION

Soil fertility can only be improved and maintained by using compost as an integral part of agriculture since the present practices deplete organic matter and continue to exhaust soils. The demand for quality compost increases day by day as compost plays a vital role in more environmentally regulated agricultural systems. Compost and composting may be the best choice as well as opportunity for added income for many livestock. However, composting may be an attractive financial alternative as well as a value-added opportunity as landfill space and openings decrease.

CONSENT FOR PUBLICATION

Not applicable.

CONFLICT OF INTEREST

The author declares no conflict of interest, financial or otherwise.

ACKNOWLEDGEMENT

Declared none.

REFERENCES

[1]   S. Yildiz, E. Olmez, and K. Alparslan, "Compost Technologies and Applications in Istanbul", *Composting Systems and Compost Application Areas Workshop,* 2009 Istanbul Turkey

[2]   E.I.A. Topal, and M. Topal, "A Review on Compost Standards", *Journal of Nevsehir Science and Technology.,* vol. 2, no. 2, pp. 85-108, 2013. [Nevsehir Turkey.].

[3]   Kafeel Ahmad, Gauhar Mahmood, and R.C. Trivedi, Kafeel Ahmad, "Municipal solid waste management in Indian cities", *A review Mufeed Sharholy a , Waste mamgement 28,* pp. 459-467, 2008.

[4]   K. Rotich, and Zhao Yongsheng, *Municipal solid waste management challenges in developing countries – Kenyan case study Waste mamgement 26.,* pp. 92-100, 2006.

[5]   M. Reyes-Torres, E.R. Oviedo-Ocaña, I. Dominguez, D. Komilis, and A. Sánchez, "A systematic review on the composting of green waste: Feedstock quality and optimization strategies", *Waste Manag.,* vol. 77, pp. 486-499, 2018.
[http://dx.doi.org/10.1016/j.wasman.2018.04.037] [PMID: 29709309]

[6]   Mufeed Sharholy, Kafeel Ahmad, R.C. Vaishya, and R.D. Gupta, *Municipal solid waste characteristics and management in Allahabad, India Waste mamgement 27.,* pp. 490-496, 2007.

[7]   M.J. Krause, G.W. Chickering, T.G. Townsend, and P. Pullammanappallil, "Effects of temperature and particle size on the biochemical methane potential of municipal solid waste components", *Waste Manag.,* vol. 71, pp. 25-30, 2018.
[http://dx.doi.org/10.1016/j.wasman.2017.11.015] [PMID: 29128251]

[8]   D.T. Sponza, and O.N. Ağdağ, "Effects of shredding of wastes on the treatment of municipal solid wastes (MSWs) in simulated anaerobic recycled reactors", *Enzyme Microb. Technol.,* vol. 36, no. 1, pp. 25-33, 2005.
[http://dx.doi.org/10.1016/j.enzmictec.2004.03.021]

[9]   A. Avcioglu, U. Turker, Z. Atasoy, and D. Kocturk, *Renewable Energies of Agricultural Origin, Biofuels.* Nobel: Ankara, Turkey, 2011.

[10]   I. Ozturk, I. Demir, M. Altinbas, O.A. Arikan, T. Ciftci, I. Cakmak, L. Ozturk, S. Yildiz, and A. Kiris, *Compost Handbook (Technical Book Series - 1).* ISTAC – TUBITAK: Istanbul, Turkey, 2015.

[11]   S. Panigrahi, and B.K. Dubey, "A critical review on operating parameters and strategies to improve the biogas yield from anaerobic digestion of organic fraction of municipal solid waste", *Renew. Energy,* vol. 143, pp. 779-797, 2019.
[http://dx.doi.org/10.1016/j.renene.2019.05.040]

[12]   K. Izumi, Y. Okishio, N. Nagao, C. Niwa, S. Yamamoto, and T. Toda, "Effects of particle size on anaerobic digestion of food waste", *Int. Biodeterior. Biodegradation,* vol. 64, no. 7, pp. 601-608, 2010.
[http://dx.doi.org/10.1016/j.ibiod.2010.06.013]

[13]   V. Ashok, *Sustainable solid waste management: An integrated approach for Asian countries Waste Management 29.,* pp. 1438-1448, 2009.

CHAPTER 6

An Application of EJSCREEN for the Examination of Environmental Justice in Metropolitan Areas of Ohio, USA

Ashok Kumar[1,*], **Lakshika Nishadhi Kuruppuarachchi**[1] and **Saisantosh Vamshi Harsha Madiraju**[1]

[1] College of Engineering, The University of Toledo, Toledo, Ohio, USA 43606

Abstract: Over the past few decades, the notion of Environmental Justice (EJ) in the United States has grown. Many empirical studies prove how low-income and minority neighborhoods are excessively exposed to environmental burdens. This chapter aims to present an approach to identifying EJ concerns facing minority and low-income populations in the metropolitan areas in Ohio by analyzing their distribution using EJSCREEN, a screening and mapping tool developed by the USEPA. Twelve metropolitan areas were considered to examine environmental and demographical information. The metropolitan areas are integrated geographic regions comprised of at least one city or urban area and adjacent communities. In assessing the demographic inequalities and environmental risk in the regions of the metropolitan areas, the EJSCREEN tool was used to generate EJ standard reports for all the zip codes in the metropolitan areas. Two-sample t-test results indicate that diesel PM, hazardous waste, RMP sites, lead paint, traffic proximity, respiratory hazard risk, and air toxic cancer risk are significantly higher in areas where a higher proportion of low-income and minority populations live than the areas with a lower proportion with low-income and minority populations. These environmental indicators are directly associated with air pollution.

Keywords: EJSCREEN, Environmental Justice, Metropolitan Areas, Ohio, Statistical Analysis.

INTRODUCTION

According to the United States Environmental Protection Agency (USEPA), "Environmental justice is the fair treatment and meaningful involvement of all people regardless of race, color, national origin, or income, concerning the development, implementation, and enforcement of environmental laws, regulations and policies" [1]. Fair treatment means every person in the community

---

** **Corresponding author Ashok Kumar:** College of Engineering, The University of Toledo, Toledo, Ohio, USA 43606; Tel: 419-934-0878; E-mail: akumar@utoledo.edu*

G. Venkatesan, S. Lakshmana Prabu and M. Rengasamy (Eds.)

have an unbiased fair share of the negative environmental consequences resulting from the policies and operations of industrial, governmental, and commercial sectors. People can participate in decision-making activities that may affect their environment/health. The contribution of the public can influence the regulatory agency's decision. The queries/concerns from the communities will be considered in the decision-making process. Decision-makers will facilitate the involvement of those potentially affected [2].

Environmental justice revolves around people residing in a city/any place in different communities. This differentiation comes with people of different races, wealth, languages, and different ethnicity [3]. High-income communities have access to nutritious organic food and are often far away from the emitting pollution-emitting freeways. However, the case is different for the low-income; even if nutritious food and other facilities are available, they are unaffordable. Minority people are exposed to industrial sites, polluted ports, highways, and sometimes hazardous waste. To run the essentials of a city, pollution is created in terms of air, water, and land. So, these differences make people breathe unhealthy air, drink polluted water, and live near toxins (Fig. **1**) [4]. In simple terms, even if all the communities live in the same city, they lead a very different life from each other.

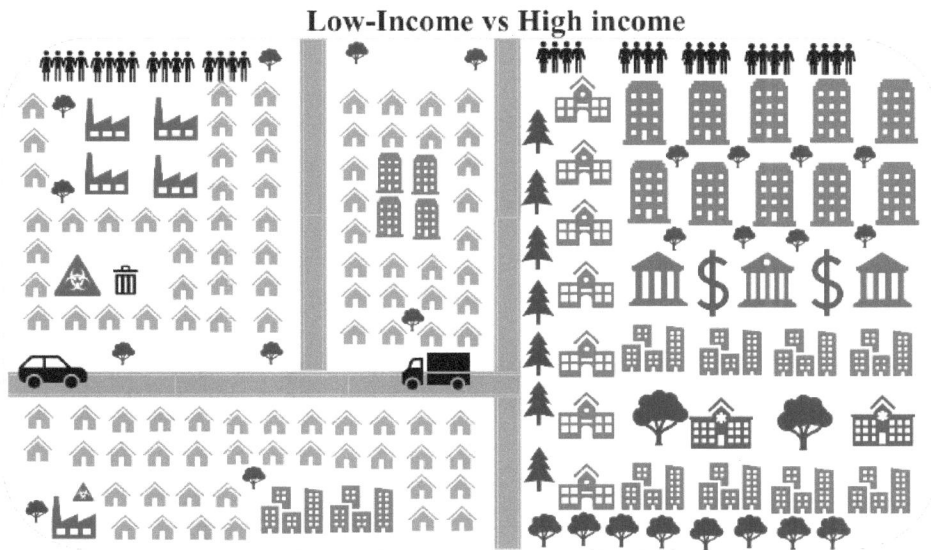

Fig. (**1**). A sample representation of a lack of environmental justice in a metropolitan area.

Most recent environmental injustices in the US have shown that vulnerable neighborhoods with a higher percentage of minority people are more likely to be exposed to more pollution, the placement of pipelines and factories, and the ongoing coronavirus disease since 2020. A growing number of proposals have been passed or are under consideration across the country that explicitly addresses the issue of "Environmental Justice." The pandemic has further highlighted the importance of addressing environmental and health inequities.

This chapter presents a simple approach to analyzing EJ issues related to an area or industry. The software developed by the USEPA is applied to get primary data for further analysis of the EJ issue. It is hoped that this chapter will motivate other environmental professionals to examine the EJ in their communities.

LITERATURE REVIEW

Generally, the facilities emitting toxic pollutants and landfills were in the places of poor, minority and low-income people reside. The people living in these communities have low meager protection from the toxic pollutants released from the facilities. Environmental justice took its origin in establishing the equalities between race, income, and nationality. There shouldn't be any bias against the poor or minority communities getting exposed to the toxins from industrial facilities. The environmental justice revolution started in the United States with the encountered incidents in 1982 (North Carolina), Warren county residents protested against dumping industrial waste in their community [5]. Later USEPA identified four landfills located in similar districts where minority people reside. In 1987, a report from the United Church of Christ stated: "the presence of hazardous wastes in racial and ethnic communities throughout the United States" [6]. In 1992, President George H.W. Bush founded the office of environmental justice in the USEPA [7]. In 1994, President Bill Clinton signed the executive order to address "Federal Actions to Address Environmental Justice in Minority Populations and Low-Income Populations." Still, it was not made a law in congress [8]. In 2001 George W. Bush shifted the focus of environmental justice from Low Income & Minority Communities to All Vulnerable People [7]. In 2009, President Barack Obama's administration supported environmental justice initiatives that assess climate change vulnerabilities and developed regional solutions, and identified innovative ways to help the most vulnerable communities to prepare for the impacts of climate change through "Environmental justice progress reports (Executive Office of the President, 2013). Former EPA Administrator Lisa P. Jackson established environmental justice as an agency-wide (USEPA) priority. The EJ 2020Action Agenda is EPA's strategic plan for advancing environmental justice for the years 2016-2020, building on the work of Plan EJ 2014. This plan consists of eight priority areas and four significant national environmental justice challenges facing the nation's minority, low-income, tribal and indigenous populations [9].

Many studies have demonstrated an association between low-income and minority populations and the presence of air pollution in a selected geographical location. Most of the assessments were performed usually for a census tract, postal code, or county, assuming the same level of exposure from polluting facilities [10]. Housing located along significant roadways is more likely to consist of low-income and minority populations living in older housing, which is more susceptible to indoor exposure to outdoor air pollution [11].

Communities worldwide have raised their voice about the environmental injustice of minority communities for decades. The very first EJ protest in the U.S.A was reported in Afton, Warren County, North Carolina, against a PCB landfill in 1982 [12]. North Carolina had selected a site for dumping toxic waste from this small African American community out of several potential sites. The protestor argued that this site was chosen since 84% of the population was African American, and it was the poorest county in the state. Followed by the Warren County protest, people in poor minority communities created groups to fight the environmental burdens. As a result of all these protests and complaints, President William J. Clinton issued The Federal Actions to address EJ in the minority and low-income populations in 1994 [13]. Since then, EJ must be an essential dimension f environmental and public health policy in North America [4].

McEntee *et al.* performed a GIS analysis on diesel particulate matter (DPM), lung cancer, and asthma incidences along major traffic corridors in MA, US. The three primary objectives of their study were to first determine whether significant highway corridors in Massachusetts have higher rates of above incidences than those that do not and secondly to implement a hot spot analysis to locate areas with high concentrations of these incidences and finally to determine whether these areas overlap with EJ neighborhoods. The analysis of EJ neighborhoods was considered in this study to draw attention to neighborhoods and communities that may experience a disproportionate burden of negative environmental consequences. The authors have followed the definition of the EJ population as a census block group whose resident's annual median household income is equal to or less than 65% of the statewide median or whose population is made up of 25% minority, foreign-born, or lacking English language proficiency. An independent sample t-test was performed to analyze the differences in mean values between corridor towns and non-corridor towns. The hot spot analysis was done using the $Gi*$ statistic with a 95% confidence interval to identify statistically significant hot spots. The results claim that higher rates of DPM exist along major highway corridors, and the mean lung cancer incidence was higher in corridor towns than in non-corridor cities. 69% of the people live within the borders of EJ neighborhoods and areas of spatially clustered high values of DPM exposure and asthma incidence [14].

Jerrett *et al*. [4] introduced a Geographic Information Systems (GIS) based on EJ analysis of particulate air pollution in Hamilton, Canada. The authors address two research questions in their study: whether EJ populations are more likely to be exposed to higher levels of particulate air pollution than others, and how sensitive the association is between the level of particulate air pollution and socioeconomic status. The authors have used GIS to perform the EJ analysis. Socioeconomic and demographic data were drawn as an indicator of EJ. The study results endorse that EJ populations are exposed to a higher level of particulate air pollution than other groups [4].

Seo *et al*. [15] conducted the environmental analysis for the goods movement system to prevent adverse disproportionate environmental effects followed by health risks on low-income and minority populations included in the regional transportation plan. Along the freight rail and major truck corridors, their special distributions were analyzed to address environmental justice implications and share changes and projected growth in Southern California [15].

Several studies have focused on how Coronavirus disease 2019 (COVID-19) has disproportionately affected racial minorities in the United States resulting in higher infection rates, hospitalization, and death [16]. Karaye *et al*. [17] assessed the global relationship between COVID-19 case count and social vulnerability in their study. An ordinary least square model was fitted using Social Vulnerability Index and COVID-19 case count data. The results show minority status and language, household composition and disability, and housing and transportation were found to predict COVID-19 case counts in U.S. counties [17].

Son *et al*. [18] examined environmental disparities regarding exposure to Concentrated Animal Feeding Operations (CAFO) using ZIP code level environmental justice matrices in North Carolina. The authors applied two scenarios: a count method on the number of CAFOs within a ZIP code and using a buffer (15 km) method on the area-weighted number of CAFOs. Their findings indicate that CAFOs are located disproportionately in communities with a higher percentage of minorities and in low-income communities [18].

In Ohio, the EPA established a partnership with community organizations to address environmental justice concerns effectively collaboration. Cleveland Clean Air Century Campaign and Earth Day Coalition's Sustainable Cleveland Partnership are the two communities with which Ohio EPA partnered to assist projects that create/increased awareness about air quality concerning issues in low-income and minority communities. Ohio EPA developed site-specific communication plans for permits in East Liverpool and neighborhoods in the Cincinnati areas. the Ohio Environmental Education Fund (OEEF) is handled by the Ohio EPA's Office of Environmental Education [19]. OEEF awards more than USD 1 million annually in grants to schools, universities, environmental advocacy groups, industry associations, nonprofit organizations, and others for, projects that involve increasing the awareness and understanding of environmental issues throughout the state of Ohio [20]. In Ohio, the public health system encompasses the Ohio Department of Health (ODH), 113 local health departments (Fig. **2**) health care providers, and public health stakeholders. The regulations in the Ohio Revised Code and the Ohio Administrative Code local Health Departments [21]. Fig. **(2)** shows the distribution of basic essential health care throughout the state of Ohio.

Fig. (2). Local Health Departments in the State of Ohio (Source: Ohio Department of Health).

Environmental Justice in Metropolitan Areas

The communities living in metropolitan areas experience both advantages and disadvantages in terms of environment based on environmental justice in that area. For a simple understanding of environmental injustice in metropolitan areas, categorize the communities in the metropolitan area into two types, Community A and Community B. Community A consists of people with high income, racial majority, politically powerful, *etc.* They have some environmental advantages like accessibility to recreational places such as parks, trails roads, green spaces, *etc.* They have less exposure to pollution and positive health benefits. Community B consists of people with low income, minorities, and the poor. These communities are exposed to pollution from landfills, waste facilities, airports, highways, energy production, food processing, manufacturing, and other industrial facilities. They have a lack of knowledge and awareness about health risks while being exposed. The price to purchase/rent a house/land in Community A is high and not affordable to the people from Community B. They have very few options to decide the place to live. Community B people have very many issues in their lives to tackle, and the concern about environmental impact in their lives is, the least bothered. This is considered an unfair distribution of the development, implementation, and enforcement of environmental laws, regulations, and policies [1]. Environmental Justice exists if the environmental burdens and benefits are fairly distributed in an area regardless of race, color, national origin, or income. The comparison between Community A and Community B is represented in (Fig. 3).

Fig. (3). A sample comparison of Community A and Community B with benign and detriments.

Several tools already exist to identify and map those metropolitan areas with potential environmental justice concerns. Kuruppuarachchi *et al.*, 2017 present a comparison of the three primary EJ tools: EJSCREEN, CalEnviroScreen 2.0, EJ Atlas, and their methodologies. The authors have identified some standard parameters across these tools in presenting Environmental Justice and identifying environmentally burdened communities, socially burdened communities, or both [22].

Ejscreen and Environmental Indicators

EJSCREEN is an Environmental Justice screening and mapping tool developed by EPA in 2015, replacing the tool EJView. The tool provides demographic and environmental information and will combine those two to generate an EJ Index for the user-selected geographic areas [24, 25]. An EJ Index represents a single environmental indicator. The eleven environmental indicators used in EJSCREEN [23, 26] are described by the USEPA as follows:

i. Particulate Matter – $PM_{2.5}$ level in the air- an annual average of micrograms per cubic meter (µg/m3).
ii. Ozone – Summer seasonal average of daily maximum 8-hour concentration in air in parts per billion (ppb).
iii. National Air Toxic Assessment (NATA) Diesel PM - Diesel PM Level in the air, µg/m3.
iv. Wastewater Discharge Indicator – Risk-Screening Environmental Indicators (RSEI) modeled Toxic Concentrations at stream segments within 500 meters, divided by distance in kilometers.
v. Proximity to Hazardous Waste Facilities - Count of hazardous waste facilities within 5 km (or nearest beyond 5 km), each divided by distance in kilometers.
vi. Proximity to Risk Management Plan sites (RMP) – Count of RMP (potential chemical accident management plan) facilities within 5 km (or nearest one beyond 5 km), each divided by distance in kilometers.
vii. Proximity to National Priority List Sites (NPL) – Count of proposed or listed NPL – also known as superfund – sites within 5 km (or nearest one beyond 5 km), each divided by distance in kilometers.
viii. Lead Paint Indicator – Percent of housing units built before 1960 indicating potential exposure to lead paint.
ix. Traffic Proximity and Volume – Count of vehicles (average annual daily traffic) on major roads within 500 meters, divided by distance in meters.
x. NATA Respiratory Hazard Index – Ratio of exposure concentration to a health-based reference concentration.
xi. NATA Air Toxic Cancer Risk - Lifetime cancer risk from inhalation of air toxics.

The eleven environmental indicators considered in EJSCREEN can be grouped into four categories based on the critical media: water, waste, air, and others, as shown in Fig. (**4**).

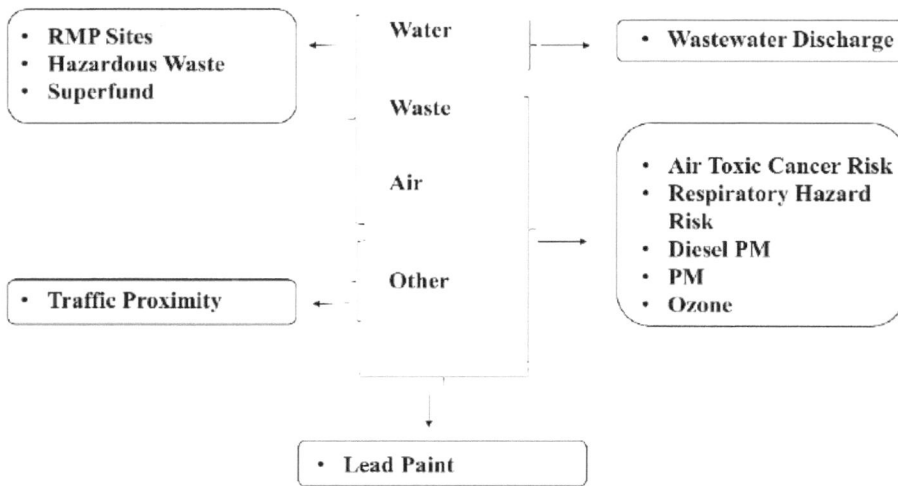

Fig. (4). Environmental indicators are categorized under critical media, water, waste, air, and others.

A census block is the smallest geographic unit, which gives the primary demographic data delineated by the United States Census Bureau for a tabulation of 100-percent data once every ten years. In EJSCREEN, the Demographic Index is calculated for each Census block group by averaging the percent low-income and percent minority [27]. The formula is calculated from the Census Bureau's American Community Survey 2008-2012 [28].

The mathematical formula for finding the Demographic Index.

$$\text{Demographic Index} = \frac{(\% \text{ minority} + \% \text{ low}-\text{income})}{2}$$

EJSCREEN users can generate reports for a selected area. These reports consist of an EJ Index as a percentile, compared with state, EPA region, and the United States EJ Index percentiles. This comparison makes it easier to understand the results of the selected area relatively.

$$\text{EJ Index} = \text{EI} * (\text{DIB} - \text{DIUS}) * \text{P}$$

where,

EI = Environmental Indicator

DIB = Demographic Index for Block Group

DIUS = Demographic Index for US

P = Population counts for block group

The EJ Index raw value itself is not reported in EJSCREEN reports. It is written in percentile terms to make the results easier to interpret. Depending on the DIB and DIUS values, the EJ Index value can be a positive or a negative number. A positive or a negative EJ Index means that the calculated DI is above the US average or below the US average, respectively [20]. However, EJ Index will be presented to the users as a percentile. The negative EJ Index will appear as lower percentiles and the positive EJ Index as higher percentiles [29].

Statistical Analysis of Ej in Ohio Metropolitan Areas

This analysis relies on the data generated using the EPA's EJSCREEN (version 2019) tool. The metropolitan areas are integrated geographic regions comprised of at least one city or urban area (with a population of at least 50,000)and adjacent communities. The estimated 2020 population (as per the US Census Bureau) of the metropolitan areas considered in this study is shown in Fig. (5).

Estimated 2020 Population of Metropolitan Areas

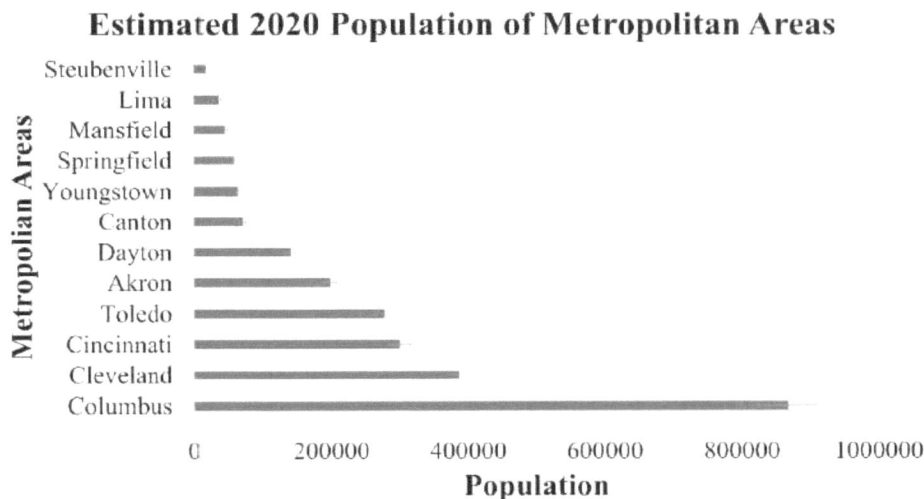

Fig. (5). Estimated Population of 12 metropolitan areas in Ohio in 2020.

In assessing the demographic inequalities and environmental risk in the metropolitan areas, the EJSCREEN tool was used in generating EJ standard reports for all the zip codes in the urban areas [30]. EJ standard report includes EJ

Indexes, a map of the selected region, environmental and demographical information used to calculate the EJ Indexes. To find out whether the environmental indicators in the low-income and minority and population areas are higher than in the other regions, a two-sample t-test was performed [31, 32]. In this case, the hypothesis stands as,

$$H_0: \mu_A \geq \mu_B \, vs \, H_1: \mu_A < \mu_B \tag{1}$$

μ_A=The mean of the environmental indicator for the areas where DI is below the USA average.

μ_B=The mean of the environmental indicator for the sites, where DI is above the US average.

According to the EJSCREEN results, the average DI index for the US is 36%.

The Two-Sample T-test

The typical comparison of two distributions is the comparison of means. The two-sample t-test was used to compare the means of random samples drawn from two populations [33].

Two-Sample t-test assumptions:

1. The data are continuous

2. The data follow the normal probability distribution.

3. The variances of the two populations are equal.

4. The two samples are independent. There is no relationship between the individuals in one-sample as compared to the other.

5. Both samples are simple random samples from their respective populations.

The individual in the population has an equal probability of being selected in the sample. Two-Sample t-test procedure for the two-sided hypothesis:

$H_0: \mu1 = \mu2 \, H_1: \mu1 \neq \mu2$,

We reject H_0 at significance level α when the computed t-statistic:

$$t = \frac{(\bar{x}_1 - \bar{x}_2) - d_0}{s_p \sqrt{\frac{1}{n_1} + \frac{1}{n_2}}}, \tag{2}$$

where,

$$s_p^2 = \frac{s_1^2(n_1-1)+s_2^2(n_2-1)}{n_1+n_2-2} \tag{3}$$

exceeds $t_{\alpha/2,n_1+n_2-2}$ or is less than $-t_{\alpha/2,n_1+n_2-2}$

One-sided alternatives suggest one-sided critical regions, as one might expect. For example,

for H₁: $\mu_1 - \mu_2 > d_0$, reject H₁: $\mu_1 - \mu_2 = d_0$ when t $> t_{\alpha,n_1+n_2-2}$

For this study, all the statistical data analysis was performed using MINITAB 16 statistical software [34], and the results are discussed in the next section.

RESULTS AND DISCUSSION

The results in Table **1** show that environmental indicators are significantly higher in areas with higher DI than the US average. Ohio metropolitan statistical areas considered in the study are presented in Fig. (**6**). The statistical analysis shows that the environmental indicators; Diesel PM, hazardous waste, RMP sites, lead paint, Traffic Proximity, Respiratory Hazard Risk, and Air Toxic Cancer Risk are significantly higher in areas where a higher proportion of low-income and minority populations live than the areas with lower proportion with low-income and minority populations. These environmental indicators are directly associated with air pollution. The National Air Toxic Assessment (NATA) is EPA's ongoing, comprehensive evaluation of air toxicity in the United States. EPA has developed the NATA to prioritize air toxic, emission sources, and location of interest for further study [35]. NATA must provide broad estimates of health risks over geographic areas of the country, not definitive risks to specific individuals or locations.

Table **2** helps to understand the correlation among each environmental indicator. The first value for each pair of indicators represents Pearson's correlation coefficient, and the second is for the *p*-value.

The correlation matrix results illustrate that respiratory hazard risk, air toxic cancer risk, traffic proximity, diesel PM, and PM are highly correlated.

Table 1. T-test results of the environmental indicators.

Demographic Index						
Environmental Indicator	**Below USA Average**		**Above USA Average**		**P-value**	**Significant (α = 0.05)**
	Mean	Std	Mean	Std		
PM	9.375	0.301	9.407	0.315	0.261	No

(Table 1) cont.....

Ozone	46.62	1.72	45.85	1.93	0.995	No
Diesel PM	0.499	0.159	0.667	0.213	0.000	Yes
Wastewater Discharge	0.140	0.587	0.148	0.519	0.468	No
Hazardous Waste	1.96	2.13	4.73	3.43	0.000	Yes
RMP Sites	0.639	0.699	1.66	1.23	0.000	Yes
Superfund	0.126	0.157	0.096	0.120	0.907	No
Lead Paint	0.437	0.216	0.652	0.177	0.000	Yes
Traffic Proximity	532	320	1032	813	0.000	Yes
Respiratory Hazard Risk	0.3637	0.0540	0.3934	0.0599	0.001	Yes
Air Toxic Cancer Risk	27.45	3.34	28.60	3.14	0.015	Yes

Fig. (6). Ohio metropolitan statistical areas– study area (Source: Ohio Department of Health).

Table 2. Correlation matrix of the environmental indicators for metropolitan areas of Ohio.

-	PM	Ozone	Diesel PM	Wastewater Discharge	Hazardous Waste	RMP Sites	Superfund	Lead Paint	Traffic Proximity	Respiratory Hazard Risk
Ozone	-.384 0.001	-	-	-	-	-	-	-	-	-

(Table 2) cont.....

Diesel PM	0.392 0.001	-0.085 0.47	-	-	-	-	-	-	-	-
Wastewater Discharge	0.193 0.099	-0.239 0.04	0.075 0.526	-	-	-	-	-	-	-
Hazardous Waste	0.459 0	-0.196 0.094	0.73 0	0.246 0.034	-	-	-	-	-	-
RMP Sites	0.17 0.148	-0.337 0.003	0.36 0.002	0.246 0.034	0.595 0	-	-	-	-	-
Superfund	-0.159 0.176	0.53 0	0.108 0.36	-0.157 0.181	-0.193 0.099	-0.274 0.018	-	-	-	-
Lead Paint	0.25 0.032	-0.412 0	0.433 0	0.026 0.827	0.417 0	0.313 0.007	-0.069 0.561	-	-	-
Traffic Proximity	0.329 0.004	-0.131 0.266	0.608 0	0.402 0	0.521 0	0.507 0	-0.033 0.781	0.191 0.102	-	-
Respiratory Hazard Risk	0.361 0.002	0.188 0.109	0.885 0	0.076 0.518	0.623 0	0.222 0.057	0.416 0	0.277 0.017	0.508 0	-
Air Toxic Cancer Risk	0.192 0.102	0.144 0.22	0.471 0	0.214 0..67	0.323 0.005	0.042 0.72	0.351 0.002	0.128 0.278	0.232 0.047	0.656 0

SUMMARY

Ohio has been recognized as one of the most air-polluted states in the USA. This study is focused to identify whether the low-income and minority populations disproportionately share this environmental consequence. The results indicate that the environmental air pollution indicators are significantly higher in the high level of low-income and minority population areas than in others. Also, results show that minority and low-income population density is higher adjacent to significant traffic and railroads. There could be many reasons for this distribution of the population. This will be a good source for future researchers to study the settlement of people along with the significant traffic and railroads or the effect of these constructions on these populations.

CONSENT FOR PUBLICATION

Not applicable.

CONFLICT OF INTEREST

The author declares no conflict of interest, financial or otherwise.

ACKNOWLEDGEMENTS

The Pollution Prevention project funded by the USEPA (PI: Dr. Kumar) motivated us to examine the issue of EJ. Any opinions, findings, conclusions, or recommendations expressed in this material are those of the author(s) and do not necessarily reflect the views of any agency. The statistical analysis was performed using Minitab version 16, statistical software. The authors have no financial interest in any software used in this study.

REFERENCES

[1]   USEPA, *Environmental justice*.https://www.epa.gov/environmentaljustice

[2]   L.A. Waller, *Environmental justice.* vol. Vol. 2. Encycl. Environmetrics, 2006.

[3]   S. Pirk, "Expanding public participation in environmental justice: methods, legislation, litigation and beyond", *J Envtl Litig,* vol. 17, p. 207, 2002.

[4]   M. Jerrett, R.T. Burnett, P. Kanaroglou, J. Eyles, N. Finkelstein, C. Giovis, and J.R. Brook, "A GIS–environmental justice analysis of particulate air pollution in Hamilton, Canada", *Environ. Plann. A,* vol. 33, no. 6, pp. 955-973, 2001.
        [http://dx.doi.org/10.1068/a33137]

[5]   B. Edwards, and A.E. Ladd, Bob Edwards, Anthony E. Ladd, "Environmental justice, swine production and farm loss in North Carolina", *Sociol. Spectr.,* vol. 20, no. 3, pp. 263-290, 2000.
        [http://dx.doi.org/10.1080/027321700405054]

[6]   United Church of Christ, *Commission for Racial Justice, Toxic wastes and race in the United States: A national report on the racial and socio-economic characteristics of communities with hazardous waste sites.* Public Data Access, 1987.

[7]   R. Holifield, "The elusive environmental justice area: Three waves of policy in the US environmental protection agency", *Environ. Justice,* vol. 5, no. 6, pp. 293-297, 2012.
        [http://dx.doi.org/10.1089/env.2012.0029]

[8]   S.L. Cutter, "Race, class and environmental justice", *Prog. Hum. Geogr.,* vol. 19, no. 1, pp. 111-122, 1995.
        [http://dx.doi.org/10.1177/030913259501900111]

[9]   US Environmental Protection Agency, *EJ 2020 action agenda: Environmental justice strategic plan.,* 2016.

[10]   M. Greenberg, "Proving environmental inequity in siting locally unwanted land uses", *Risk,* vol. 4, p. 235, 1993.

[11]   S. Saksena, P.B. Singh, R.K. Prasad, R. Prasad, P. Malhotra, V. Joshi, and R.S. Patil, "Exposure of infants to outdoor and indoor air pollution in low-income urban areas — a case study of Delhi", *J. Expo. Sci. Environ. Epidemiol.,* vol. 13, no. 3, pp. 219-230, 2003.
        [http://dx.doi.org/10.1038/sj.jea.7500273] [PMID: 12743616]

[12]   R.D. Bullard, and J. Lewis, *Environmental justice and communities of color.* San Franc, 1996.

[13]   E. Order, *Federal actions to address environmental justice in minority populations and low-income populations.,* 1994.

[14]   J.C. McEntee, and Y. Ogneva-Himmelberger, "Diesel particulate matter, lung cancer, and asthma incidences along major traffic corridors in MA, USA: A GIS analysis", *Health Place,* vol. 14, no. 4, pp. 817-828, 2008.
        [http://dx.doi.org/10.1016/j.healthplace.2008.01.002] [PMID: 18280198]

[15]   J. H. Seo, F. Wen, J. Minjares, and S. Choi, *Environmental justice analysis of minority and low-income populations adjacent to goods movement corridors in Southern California.,* 2012.

[16]   H. Schmidt, L.O. Gostin, and M.A. Williams, "Is it lawful and ethical to prioritize racial minorities for

COVID-19 vaccines?", *JAMA,* vol. 324, no. 20, pp. 2023-2024, 2020.
[http://dx.doi.org/10.1001/jama.2020.20571] [PMID: 33052391]

[17]   I.M. Karaye, and J.A. Horney, "The impact of social vulnerability on COVID-19 in the US: an analysis of spatially varying relationships", *Am. J. Prev. Med.,* vol. 59, no. 3, pp. 317-325, 2020.
[http://dx.doi.org/10.1016/j.amepre.2020.06.006] [PMID: 32703701]

[18]   J.Y. Son, R.L. Muenich, D. Schaffer-Smith, M.L. Miranda, and M.L. Bell, "Distribution of environmental justice metrics for exposure to CAFOs in North Carolina, USA", *Environ. Res.,* vol. 195, p. 110862, 2021.
[http://dx.doi.org/10.1016/j.envres.2021.110862] [PMID: 33581087]

[19]   E.P.A.   Ohio.   *Ohio   EPA   and   environmental   justice,* 2008.http://ohioepa.custhelp.com/app/answers/detail/a_id/1097/~/ohio-epa-and-environmental-justice

[20]   Office of Environmental Education, *Ohio Environmental Education Fund Mini Grant Program Guidelines 2019,* 2019.http://epa.ohio.gov/oee/EnvironmentalEducation.aspx

[21]   Ohio Department of Health, *Local public health.*https://odh.ohio.gov/wps/portal/gov/odh/about-us/local-health-departments

[22]   L.N. Kuruppuarachchi, A. Kumar, and M. Franchetti, ""A comparison of major environmental justice screening and mapping tools," Env", *Environmental Management and Sustainable Development,* vol. 6, no. 1, pp. 59-71, 2017.
[http://dx.doi.org/10.5296/emsd.v6i1.10914]

[23]   E. Ejscreen, *Environmental justice screening and mapping tool.,* 2015.

[24]   D.W. Case, "The role of information in environmental justice", *Miss. Law J.,* vol. 81, p. 701, 2011.

[25]   J. Ikeme, "Equity, environmental justice and sustainability: incomplete approaches in climate change politics", *Glob. Environ. Change,* vol. 13, no. 3, pp. 195-206, 2003.
[http://dx.doi.org/10.1016/S0959-3780(03)00047-5]

[26]   K. Cason, Y. Sanchez, and E. I. Mentor, *EJSCREEN community summary for the Montrose-Del Amo superfund sites.,* 2016.

[27]   J. Iceland, and E. Steinmetz, *The effects of using census block groups instead of census tracts when examining residential housing patterns.* vol. Vol. 5. Bur. Census, 2003.

[28]   Minnesota Population Center, "US Census Bureau's American community survey, 2008-2012", *Natl. Hist. Geogr. Inf. Syst. Version.,* vol. 110, 2012.

[29]   C. Reczek, "Sexual-and gender-minority families: A 2010 to 2020 decade in review", *J. Marriage Fam.,* vol. 82, no. 1, pp. 300-325, 2020.
[http://dx.doi.org/10.1111/jomf.12607] [PMID: 33273747]

[30]   S. Morrison, F.M. Fordyce, and E.M. Scott, "An initial assessment of spatial relationships between respiratory cases, soil metal content, air quality and deprivation indicators in Glasgow, Scotland, UK: relevance to the environmental justice agenda", *Environ. Geochem. Health,* vol. 36, no. 2, pp. 319-332, 2014.
[http://dx.doi.org/10.1007/s10653-013-9565-4] [PMID: 24203260]

[31]   US Environmental Protection Agency, *EJSCREEN: environmental justice screening and mapping tool.,* 2018.

[32]   G.S. Mills, and K.S. Neuhauser, "Quantitative methods for environmental justice assessment of transportation", *Risk Anal.,* vol. 20, no. 3, pp. 377-384, 2000.
[http://dx.doi.org/10.1111/0272-4332.203036] [PMID: 10949416]

[33]   E.S. Metzger, and J.M. Lendvay, "COMMENTARY: seeking environmental justice through public participation: a community-based water quality assessment in Bayview Hunters Point", *Environ. Pract.,* vol. 8, no. 2, pp. 104-114, 2006.
[http://dx.doi.org/10.1017/S1466046606060133]

[34] M. Version, *16, 2010, Minitab.*

[35] J.M. Logue, M.J. Small, and A.L. Robinson, "Evaluating the national air toxics assessment (NATA): Comparison of predicted and measured air toxics concentrations, risks, and sources in Pittsburgh, Pennsylvania", *Atmos. Environ.,* vol. 45, no. 2, pp. 476-484, 2011. [http://dx.doi.org/10.1016/j.atmosenv.2010.09.053]

CHAPTER 7

Comparative Adsorption Study of Acid Violet 7 and Brilliant Green Dyes in Aqueous Media using Rice Husk Ash (RHA) and Coal Fly Ash (CFA) Mixture

Irvan Dahlan[1,*], Sariyah Mahdzir[2], Andi Mulkan[3] and Haider M. Zwain[4]

[1] *School of Chemical Engineering, Universiti Sains Malaysia, Engineering Campus 14300 Nibong Tebal, Pulau Pinang, Malaysia*

[2] *School of Civil Engineering, Universiti Sains Malaysia, Engineering Campus 14300 Nibong Tebal, Pulau Pinang, Malaysia*

[3] *Mechanical Engineering Study Program, Faculty of Engineering, University of Iskandar Muda, Jalan Kampus Unida - Surien, Banda Aceh 23234, Indonesia*

[4] *College of Water Resources Engineering, Al-Qasim Green University, Al-Qasim Province, Babylon, Iraq*

Abstract: One of the concerns in wastewater pollution is the presence of colored compounds, such as dyes. Acid violet 7 (AV7) and brilliant green (BG) are examples of synthetic dyes that have been used in various applications. In this work, a comparison of AV7 and BG dye adsorption was investigated using an adsorbent prepared from the mixture of rice husk ash (RHA) and coal fly ash (CFA). The attention was focused on the major batch adsorption parameters, which include adsorbent dosage, initial dye concentration, contact time, pH, shaking speed, and temperature. A lesser amount of RHA-CFA adsorbent was found to be used for adsorbing the same concentration of BG as compared to AV7. In contrast to AV7, the adsorption of BG rapidly attained equilibrium. The effective pH for BG removal is in the pH range of 6–8, while the highest AV7 removal was obtained at a low pH value. The adsorption removal for AV7 and BG increases with rising shaking speed and temperature. Scanning electron morphology (SEM) analysis showed the morphological porous structure on the RHA–CFA adsorbent surface. X-ray diffraction (XRD) analysis indicated the presence of complex compounds containing cristobalite, quartz, and mullite compounds in the RHA–CFA adsorbent. The study revealed that RHA–CFA adsorbents can remove AV7 and BG from an aqueous medium.

Keywords: Acid violet 7 (AV7) dye, Adsorbent, Brilliant green dye, Coal fly ash, Rice husk ash.

---

* **Corresponding author Irvan Dahlan:** School of Chemical Engineering, Universiti Sains Malaysia, Engineering Campus 14300 Nibong Tebal, Pulau Pinang, Malaysia; Tel: +604-5996463; E-mail: chirvan@usm.my

G. Venkatesan, S. Lakshmana Prabu and M. Rengasamy (Eds.)

INTRODUCTION

Population growth and the development of many types of industries have led to different kinds of environmental pollutants. One of the largest environmental problems encountered by cosmetics, dyestuff, textile, and related industries is water pollution attributed to organic dye usage [1 - 3]. More than half of the commercial synthetic organic dyes come from textile-related industries, and approximately 100 tons of effluent from these industries are discharged yearly [4].

Acid violet 7 (AV7) and brilliant green (BG) dyes are among the commercial synthetic dyes that have been widely used. AV7 dye belongs to the azo group and comprises aromatic, amino, and sulfonic groups in their structure, causing higher recalcitrant and genotoxicity compared with other azo dyes [5, 6]. Fabbri *et al.* [7] stated that the structures not adjacent to the azo bond showed high reactivity. Theoretically, the increasing number of azo groups lowers the decolorization capabilities of dyes. The usage of azo dyes in textile dyeing is extensive because of their superior fastness on the applied fabric, high photolytic stability, and resistance to microbial degradation [8]. Meanwhile, BG dye is categorized as a basic dye because it involves a high brilliance and intensity of colors and is highly visible even at a remarkably low concentration [9]. BG is a sulfate of di-(--diethylamino) triphenyl carbonyl anhydride [10] that presents a complex chemical structure. Both dyes are associated with a stable and complex structure and are simultaneously linked to carcinogenic, mutagenic, and neurotoxicity characteristics. Thus, the removal of these dyes is necessary because they can threaten water organisms, whereas their redundancy can alter the native color of the receiving water.

Hence, dye-containing wastewater must be treated before it is released into receiving water to prevent the aforementioned consequence. Many technologies have been developed for the treatment of dye-containing wastewater. Adsorption is among the methods that has been adopted for AV7 and BG dye removal. Previous studies showed that activated carbon is the best adsorbent. Regardless of their wide use for the removal of dye-containing wastewater, exploration of low-cost adsorbents prepared from agricultural and industrial wastes has been widely investigated for activated carbon substitution.

In Malaysia, rice husk ash (RHA) and coal fly ash (CFA) are among the examples of abundantly obtainable solid wastes generated from the combustion process at high temperatures of rice husk and coal in the rice mill and coal-fired power plant industries, respectively. The RHA was collected from the dust collection device attached upstream to the stacks of rice husk-fired boilers and furnaces [11]. Rao *et al.* [12] stated that approximately 22% of husk is produced from 1000 kg of paddy and milled rice and approximately 25% of husk weight is converted into ash during the combustion process. More than 43 million tons of RHA are expected to be generated worldwide based on the aforementioned conversion and paddy rice production statistics in 2018 [13]. Meanwhile, Malaysia alone generates almost 150 thousand tons of RHA. The disposal in landfills or open fields can be problematic and may cause serious environmental and human-health related problems due to the low bulk density of RHA [14]. Concurrently, coal-fired power plants in Malaysia have produced a considerable amount of coal fly ash (CFA) of approximately 6.8 and 1.7 million tons of coal bottom ash annually [15].

RHA and CFA have been utilized in many applications and possess satisfactory adsorption capability in removing many types of dyes. However, almost all investigations only adopted one form of ash, either RHA or CFA. The effect of the preparation of RHA–CFA mixture was recently addressed by using three different methods (*i.e.*, reflux, magnetic co-precipitation, and magnetic template) for the removal of AV7 and BG dyes [16]. The optimum RHA–CFA adsorbent preparation condition was also investigated by using response surface methodology (RSM) and artificial neural network (ANN). Current studies showed that 2nd-order RSM and ANN models can be applied to predict and optimize the efficiency of RHA–CFA adsorbent toward dye-containing wastewater [17]. In continuation of the current study, investigating the influence of various parameters affecting RHA–CFA adsorbents is important. Parameters, such as pH, the initial concentration of solute and sorbent, agitation speed, temperature, and contact time, were usually studied in batch adsorption.

This study is crucial due to the absence of available information regarding the effects of various parameters for this kind of adsorbent during batch AV7 and BG dye adsorption. The current study also helps define the performance of RHA–CFA adsorbents. Furthermore, the extent of AV7 and BG dye removal significantly varies for different types of adsorbents, and the effects of various parameters reported in the literature are also inconsistent. Hence, performing a thorough study on this issue is useful. Thus, the effects of adsorbent dosage, initial dye concentration, contact time, pH, shaking speed, and temperature on the adsorption efficiency of RHA–CFA adsorbents toward dye were studied in the present work. The adsorption efficiency of RHA–CFA adsorbents for the removal of AV7 and BG dyes from aqueous solutions in a batch process is also quantified and compared. In addition, the prepared and spent RHA–CFA adsorbents were characterized by X-ray fluorescence (XRF), scanning electron microscopy (SEM), Fourier transform infrared (FTIR), and X-ray diffraction (XRD) to assess changes due to AV7 and BG dye adsorption.

MATERIALS AND METHODS

Materials

The RHA in the form of black coarse dry ash was received from Kilang Beras & Minyak Sin Guan Hup Sdn. Bhd. Nibong Tebal, Penang. CFA in the form of brownish powder was collected from Kapar Coal-Fired Power Plant, Selangor. AV7 and BG dyes were bought from Sigma-Aldrich, Malaysia. Analytical grade of other chemicals was used in this work.

Synthesis of RHA–CFA Adsorbents

In this study, the RHA–CFA adsorbents were prepared following the reflux method proposed by Ai *et al.* [18] with few alterations and according to the optimum condition obtained from previous studies [16, 17]. The adsorbent was prepared with a 3:1 ratio of RHA to CFA. The mixture was then mixed with 1.21 M of sodium hydroxide (NaOH) solution and stirred at 80 °C for 2 h. Afterward, the mixture was washed repeatedly with deionized water using the filtration method until the filtrate was neutralized (around pH 7). The adsorbent was then dried at 110 °C overnight and was subsequently used as adsorbent for the batch adsorption experiments.

Batch Adsorption Studies

The batch experiments were performed in accordance with the previous standard procedures [19, 20]. In this procedure, dye concentration was prepared from the dilution of the stock dye solution. The adsorption experiment was conducted by adding a specific amount of RHA–CFA adsorbent into a 250 mL conical flask filled with 100 mL of either AV7 or BG dye. Together with a blank sample, the conical flasks were wrapped and shaken by using Sartorius (Ceriomat® SII) shaker at a certain parameter condition. The dye concentration before and after adsorption was determined by using DR 2500 spectrophotometer (Shimadzu, Japan) at wavelengths of 520 nm [21] and 623 nm [22] for AV7 and BG, respectively. Triplicate experiment was employed for every experiment and only average value was reported. The dye removal efficiency was obtained from the following equation:

$$\text{Removal efficiency (\%)} = \frac{C_0 - C_e}{C_0} \times 100 \tag{1}$$

where C_0 (mg/L) is an initial dye concentration, and C_e (mg/L) is the dye concentration after adsorption.

In this batch adsorption experiment, parameters that affect the adsorption of dyes by the prepared RHA–CFA adsorbent, which include adsorbent dosage, the initial concentration of synthetic dyes, contact time, pH of the solution, agitation or shaking speed, and operational temperature, were studied. During the experiment, one factor is varied while the other factors were fixed. The effects of adsorbent dosage were conducted by adding a different amount of adsorbent in 100 mL of fixed initial dye concentration (*i.e.*, 100, 200, and 300 mg/L). The amount of adsorbent used for AV7 and BG adsorption is in the range of 0.1–1.4 and 0.1–0.6 g, respectively. The effects of initial dye concentration and contact time were studied for 150 min. Dye solutions with different pH (ranging from pH 2 to pH 10) were used in the adsorption experimental process. The experiment also included the actual pH of dyes, which were 6.8 and 3.6 for AV7 and BG solution, respectively. The pH of the solution was adjusted using 0.1 M NaOH and 0.1 M HCl to obtain the desired pH. Different shaking speeds ranging from 100 rpm to 300 rpm were studied. In addition, the operational adsorption temperature of different dye concentrations (200, 250, and 300 mg/L) of dye solutions was conducted at 30 °C, 40 °C, 50 °C, and 60 °C.

Characterization of Adsorbent

The XRF analysis was conducted on RHA, CFA, and prepared adsorbents using Rigaku RIX 3000 to determine the available chemical compounds. SEM analysis was conducted on prepared and spent adsorbents using Zeiss Supra 35VP to study the surface morphology and verify the presence of porosity on the surface of the adsorbents. The surface functional groups of the prepared and spent adsorbents were detected and determined by using FTIR spectrometer (Shimadzu IR Prestige-21). Meanwhile, the XRD analysis was performed using Bruker AXS Diffractometer D8 at the scanning range from 10° to 90° in 2θ scale. The results were analyzed using EVA software to determine the present crystalline phases in the adsorbents.

RESULTS AND DISCUSSION

Effect of Dosage

Adsorbent dosage is an important parameter used to study the adsorption capacity at a certain concentration. The effects of adsorbent dosage at different initial dye concentrations are shown in Fig. (**1**). Obtained results reveal that the percentage removal of both dyes increases with rising amount of adsorbent loading. This finding was due to the increasing number of total surface area and the availability of the adsorption sites as the amount of adsorbent was raised [19].

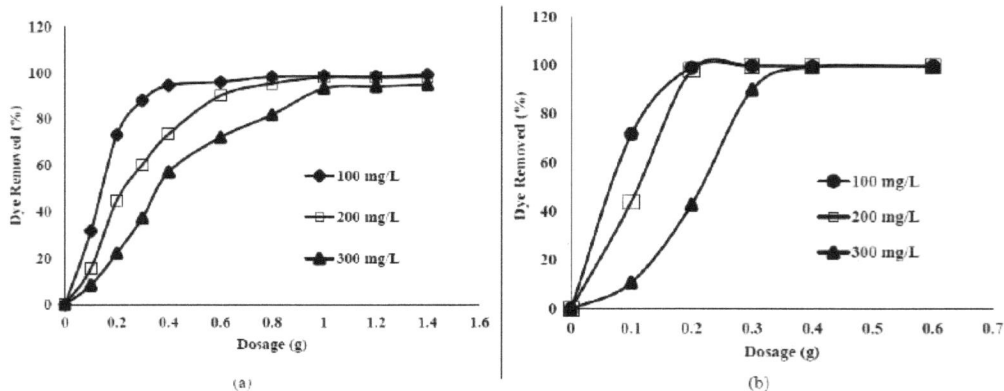

Fig. (1). Effect of adsorbent dose on the adsorption of (a) AV7 and (b) BG dyes.

For the removal of AV7 (Fig. **1a**), the increment of dye removal becomes constant at dosages 0.4, 0.8, and 1.0 g for the initial concentrations of 100, 200, and 300 mg/L, respectively. This result indicates that a high amount of adsorbent is needed for AV7 adsorption as the initial concentration increases. Similar trends were also observed in the adsorption of BG dye (Fig. **1b**), where a high amount of adsorbent is necessary for adsorption of additional concentrated BG solution. However, compared with AV7, a less amount of adsorbent was used for adsorbing the same BG concentration. Thus, 0.2 g of adsorbent is needed for the adsorption of 100 and 200 mg/L, while 0.4 g of adsorbent is necessary for the adsorption of 300 mg/L BG dye solution. This result shows that the prepared adsorbent can remove BG dye more effectively compared with AV7 dye. This finding is due to the higher solubility of AV7 dye (solubility: 120 g/L) than BG dye (solubility: 40 g/L), which leads to the higher attraction of AV7 molecules to the water than that to the adsorbent, leading to a low percentage of AV7 removal [23].

Effect of Initial Concentration and Contact Time

The influences of the initial concentration of dyes were studied at concentrations of 100, 200, and 300 mg/L. The effect of contact time was also studied, and all the results are presented in Fig. (**2**). The experiment was performed at 30 °C with a shaking speed of 150 rpm for 150 min. All the graphs from both figures show smooth and continuous curves that lead to saturation, suggesting the possible monolayer coverage of dye on the surface of the adsorbent [24 - 26]. For AV7 dye removal shown in Fig. (**2a**), the maximum removal percentages for adsorption of 100, 200, and 300 mg/L of dyes are 71.99%, 47.92%, and 29.34% before attaining equilibrium at 90, 60, and 30 min, respectively. As the initial concentration was increased, the percentage removal of dye was decreased, and the time needed for the reaction to attain equilibrium was reduced. Increasing initial dye concentrations eventually raises the competition between AV7 molecules to be bound onto the adsorbent surface, causing the adsorbent bonding sites to establish saturation rapidly.

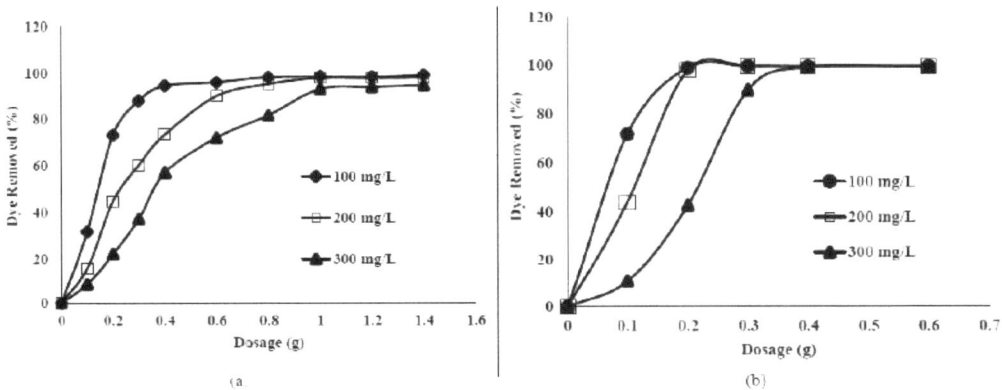

Fig. (2). Dyes adsorbed at different initial concentrations and time for (a) AV7 and (b) BG.

The adsorption of BG dye attained equilibrium faster than that of AV7 dye. Fig. (**2b**) shows that the adsorption process becomes gradual after 60 min for the initial concentration of 100 mg/L with a percentage removal of 92.2%. Meanwhile, for the initial BG concentration of 200 and 300 mg/L, both reactions attained equilibrium after 10 min of experiment with removal percentages of 67.29% and 44.29%, respectively. Similar to the adsorption of AV7, the removal of BG is decreasing, and less time is needed for the reaction to obtain equilibrium when the initial concentration of dyes was increased.

Effect of pH

In adsorption process, the pH of the solution can affect the adsorption capacity by changing the adsorbent surface charge and degree of ionization compounds present in the solution, which leads to the changes in dye structural stability as well as its color intensity [22]. In this study, the effects of pH on the BG and AV7 dye adsorption were investigated at varied pH ranging from 2 to 10 using certain adsorbent dosages (*i.e.*, 0.2 g and 0.1 g for AV7 and BG removal, respectively) and 100 mg/L of adsorbate for 60 min at 150 rpm and 30 °C. The solutions with the actual pH of both dyes were included in the observation as a blank solution. The actual pH for AV7 dye solution is 6.8 while that for BG dye is 3.6. The results of these studies are presented in Fig. (**3**).

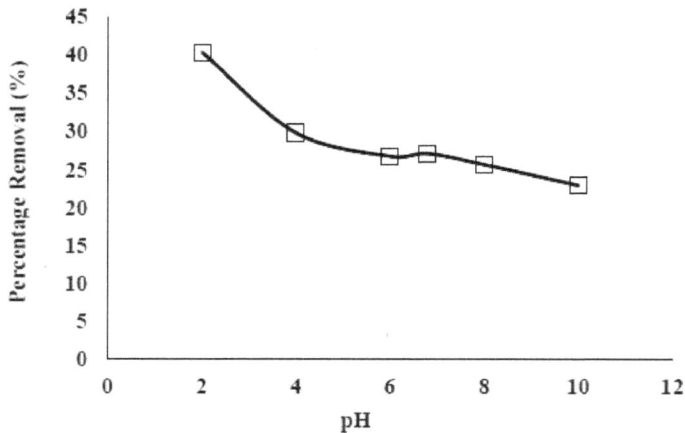

(a)

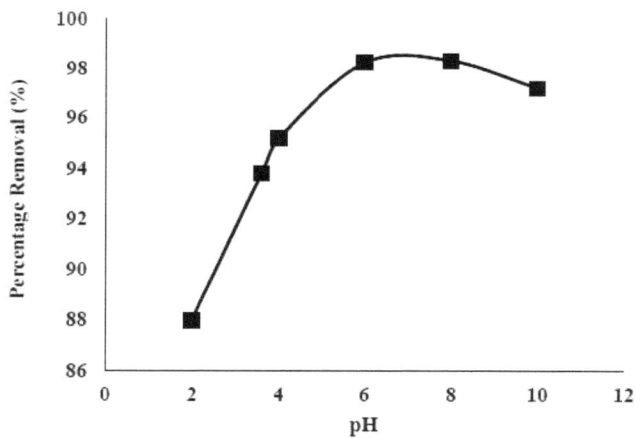

(b)

Fig. (3). Dyes adsorbed at different pH for (a) AV7 and (b) BG.

Fig. (**3a**) shows that the removal of AV7 dye was gradually decreased as the pH of the adsorbate increased. This phenomenon occurred due to the characteristics of adsorbent and dye molecules that carry negative charges. The RHA–CFA adsorbent acquires negative charge surrounding the surface during the adsorbent preparation process, while AV7 is a negatively charged anionic dye. This condition leads to the static repulsion of both adsorbent and dye molecules, which decreased the adsorption efficiency as the pH of solution is increased.

On the contrary, Fig. (**3b**) shows that the effective pH for BG removal is in the range of pH 6 and 8. The graph also indicates that the adsorption of BG dye increased by raising the pH of the solution up to a certain point and started to decrease afterward. BG, one of the cationic dyes, carries a positive charge. The prepared RHA–CFA adsorbent carries a negative charge, thus attracting the positive charge of BG dye molecules and binding them together, which leads to the high adsorption efficiency when the pH was increased. However, a sudden decrease in BG removal after a certain point (pH 8) was observed due to the point of zero charge (pH_{PZC}) of the adsorbent. Point of zero charge is defined as the pH value where one or more components of surface charge vanish at a specific temperature, pressure, and aqueous solution composition. At pH <pHpzc, adsorbent surface has a net positive charge whereas that at pH >pHpzc has a net negative charge [27]. Similarly, Nandi *et al*. [19] suggested that above the pH_{PZC}, the adsorbent acquires negative charge, leading to decreased dye uptake due to the neutralization of dye molecules at this pH.

Effect of Shaking Speed

Shaking or agitation speed is an important parameter in adsorption because it can influence the distribution of the solute in the bulk solution and the formation of the external boundary film [19]. Fig. (**4**) shows the percentage removal of AV7 and BG dyes at different shaking speeds (100, 150, 200, 250, and 300 rpm) within the contact time of 180 min. From both figures, the adsorption of AV7 and BG dyes increased with rising shaking speed from 100 rpm to 200 rpm. The increasing shaking speed induces the reduction in film boundary layer surrounding particles, which leads to rising external film transfer coefficient, thus leading to high dye removal [19]. In addition, the graphs revealed that minimal contact time is needed for the adsorption process to attain equilibrium as the shaking speed increased. Increased collision occurrences between the adsorption sites of the adsorbent and dyes molecules can also be a reason for the rising dye adsorption capacity at high shaking rates [28].

Increasing the shaking speed by up to 300 rpm shows no distinct results. Kabbashi *et al*. [29] suggested that as the speed increases, the suspension may not become homogeneous due to the rapid agitation. This phenomenon will increase the boundary layer between the solid and liquid phases and lead to the equilibrium phase. Therefore, the adsorption of AV7 and BG dyes achieves the maximum value at a shaking speed of 200 rpm.

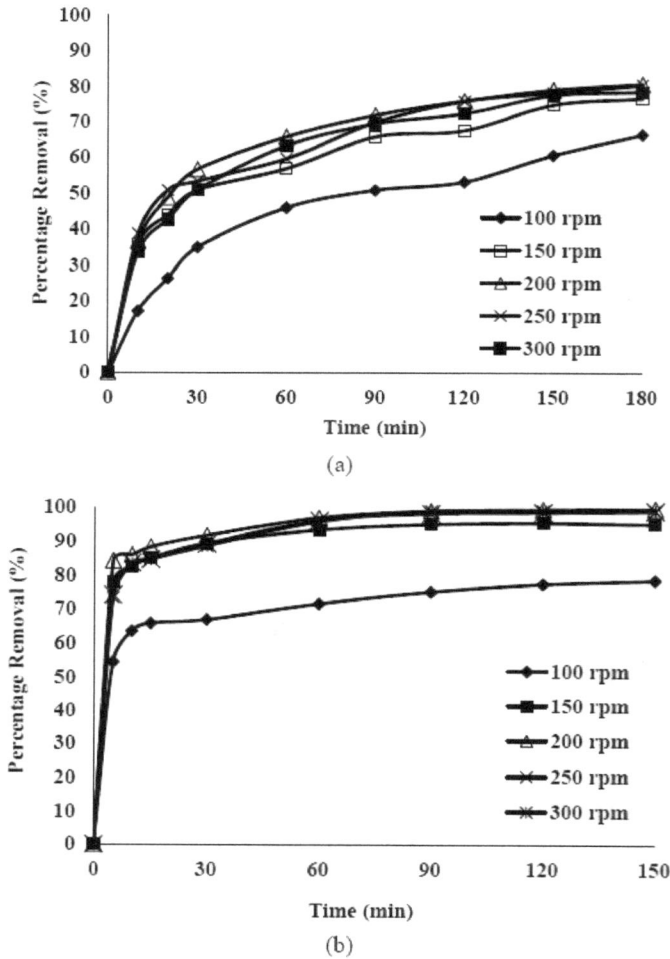

Fig. (4). Dyes adsorbed at different shaking speed for (a) AV7 and (b) BG.

Effect of Temperature

Temperature often plays a significant role in the adsorption process. In this study, the experiments were conducted at four different temperatures (*i.e.*, 30 °C, 40 °C, 50 °C, and 60 °C). The percentage removal of AV7 and BG dyes at different temperatures is shown in Fig. (**5**). The experimental results revealed that AV7 and BG dyes have similar trends, demonstrating that the percentage removal of both dyes increases with rising temperature. The possible reason for this observation is explained by the increased mobility of dye ion in temperature to have sufficient energy to undergo an interaction with the active sites on the adsorbent surface [30]. This phenomenon is due to the possibility of increased porosity and total pore volume of the adsorbent as the temperature rises. High temperatures can also induce enlargement of pore size and swelling on the internal structure of the adsorbent, which leads to the penetration of large dye molecules [31, 32]. The increasing removal percentage with rising temperature shows that the reaction during the adsorption is an endothermic process [33] for AV7 and BG dyes.

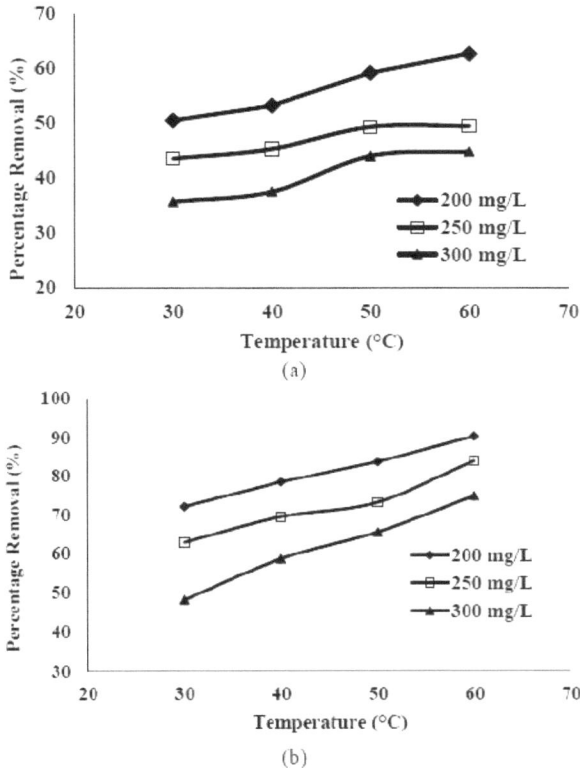

Fig. (5). Dyes adsorbed at different temperature for (a) AV7 and (b) BG.

Characterization of the RHA–CFA Adsorbent

X-Ray Fluorescent (XRF)

The chemical composition of the raw materials (RHA and CFA) and prepared RHA–CFA adsorbent were analyzed using an XRF. This analysis revealed that the RHA comprised the following: 70.58% SiO_2, 6.85% Fe_2O_3, 6.67% Al_2O_3, 5.1% C, 3.34% CaO, 2.46% Cr_2O_3, 1.78% K_2O, 0.85% MgO, 0.85% P_2O, and 1.52% others. The CFA comprised the following: 49.5% SiO_2, 20.11% Al_2O_3, 11.29% Fe_2O_3, 10.58% CaO, 2.49% C, 1.79% MgO, 1.58% K_2O, 0.82% TiO_2, and 1.78% others. Meanwhile, the prepared RHA–CFA adsorbent comprised 57.78%

SiO_2, 10.5% C, 9.10% Al_2O_3, 8.97% Fe_2O_3, 6.18% CaO, 1.81% Cr_2O_3, 1.54% MgO, 1.25% K_2O, 0.79% P_2O, and 2.08% others.

Surface Morphology

The morphological analysis of the adsorbents was conducted using SEM. Determining the particle shape, porosity, and appropriate size distribution of the adsorbent is useful [34]. The analysis was examined at the magnification of 200X to 3.0KX for clear observation. The previous study showed that raw RHA has the shape of a corn-like structure with the porous internal structure of honeycomb [35], while CFA is a hollow smooth spherical particle [36, 37]. Figs. (**6 - 8**) show the SEM images of the prepared and spent RHA–CFA adsorbents, revealing their surface texture and porosity. From the images, the presence of RHA and CFA in the adsorbents can still be observed. For prepared RHA–CFA adsorbents, Fig. (**6**) shows the attachment of some RHA and CFA particles to each other, which is attributed to the preparation process, thus forming large or complex particles. However, these combined compounds can lower the total surface area, thus decreasing the adsorption capacity of the adsorbents. The presence of micro-size particles can also be seen attached on the surface of CFA and the internal surface of RHA. These micro-size particles were formed during the preparation process. The images of spent RHA–CFA adsorbents after adsorption of AV7 and BG dyes are shown in Figs. (**7 and 8**), respectively. The images of spent adsorbents show that surface morphologies comprise irregular rough particles with different shapes and sizes, which are most probably due to the agglomeration of molecules of dyes inside and around the RHA and CFA particles.

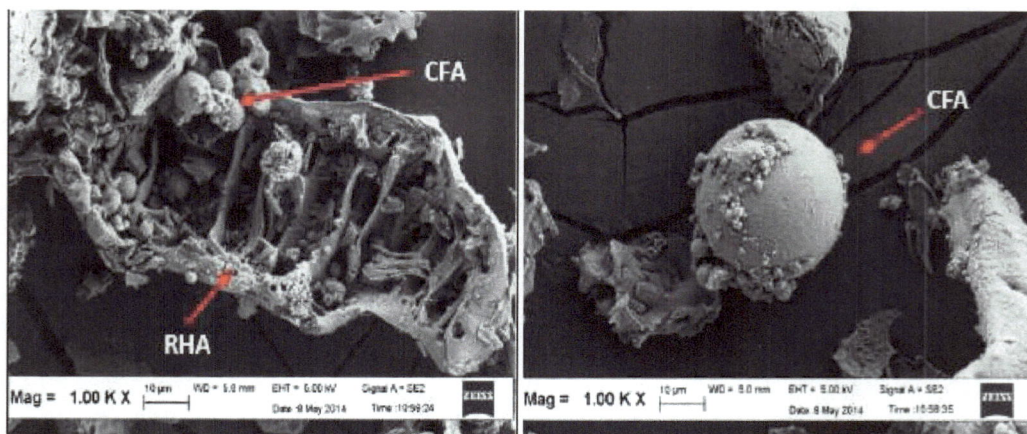

Fig. (6). SEM images of prepared RHA-CFA adsorbent.

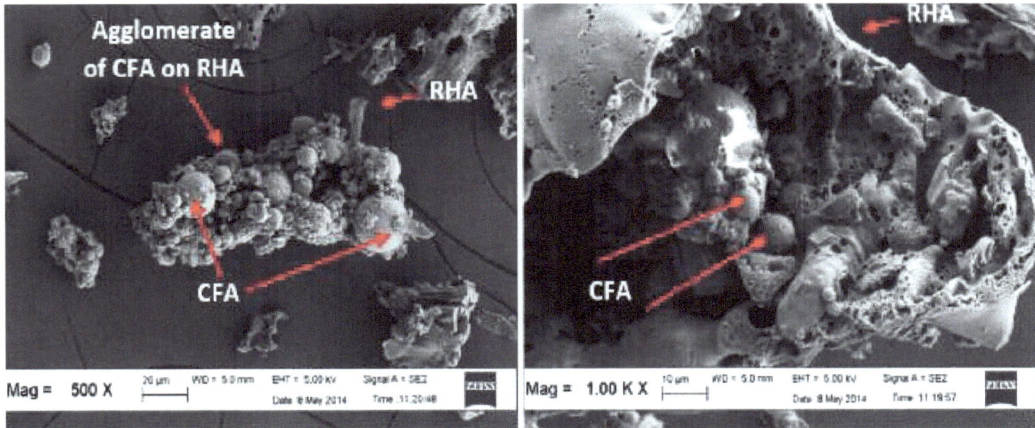

Fig. (7). SEM images of spent RHA-CFA adsorbent for BG adsorption.

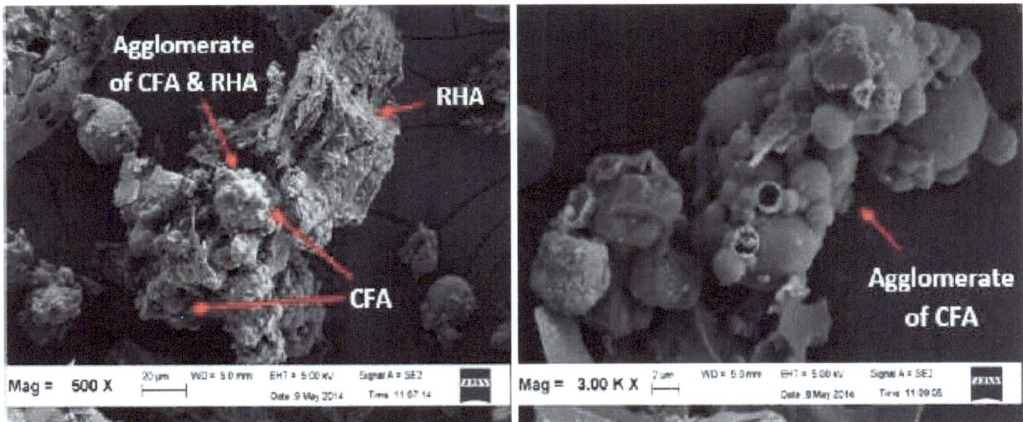

Fig. (8). SEM images of spent RHA-CFA adsorbent for AV7 adsorption.

Fourier Transform Infrared (FTIR)

FTIR is one of the methods used to characterize the adsorbent by determining the functional groups present on the surface of the adsorbent. Identifying the available functional groups will help in studying the reaction that occurs during the adsorption process. Fig. (**9**) shows the FTIR results for adsorbents for AV7 and BG dye adsorption. Fig. (**9**) reveals that the prepared RHA–CFA adsorbent has a broad stretching band with a peak at 3444 cm^{-1}. This band is predicted to be a silanol (SiOH), one of the silica-based derivatives. The silanols located at the surface of the adsorbent raise the broad peak of the O-H band that range from 3200– 3500 cm^{-1} to a wavenumber range of 3200–3700 cm^{-1}. After AV7 and BG dye adsorption, this peak shows low intensity with two bands at 3417 and 3452 cm^{-1}, suggesting the presence of primary amine (NH_2) asymmetric stretch attached on aromatic compounds. Moreover, the fading of peaks around the wavenumber of 3760 cm^{-1} signified the reduction in the amount of O-H stretch silanol group on prepared adsorbents [38].

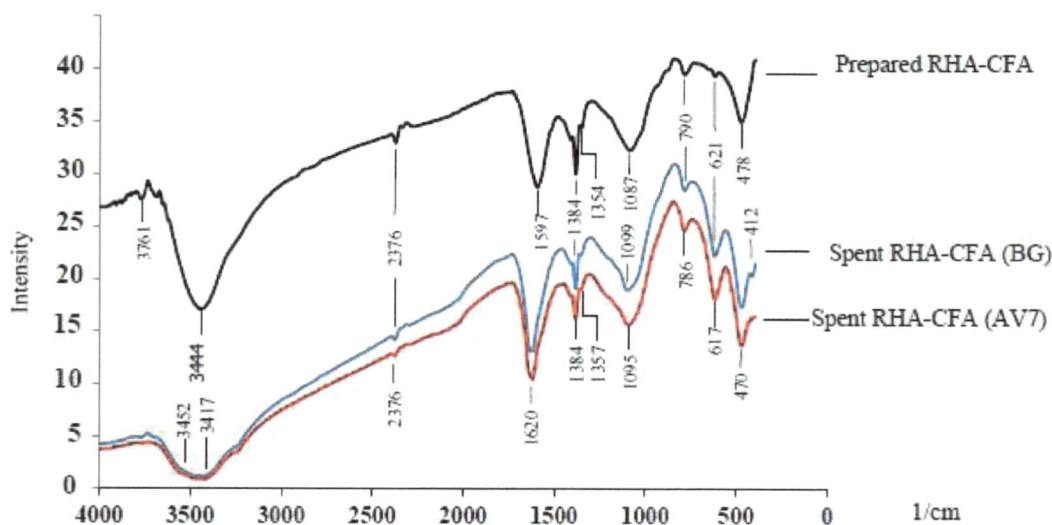

Fig. (9). FTIR analysis for prepared and spent RHA-CFA adsorbents.

The peak at frequency 2376 cm^{-1} is predicted to be silane functional group. Silane (Si-H) is a functional group that has only hydrogen atom attached to the silicon. Moreover, the presence of an intense Si-O-Si asymmetric stretch at 1087 cm^{-1} confirms the presence of silica compound in the prepared adsorbent. This peak is slightly shifted to the left at 1095 and 1099 cm^{-1} for both spent adsorbents used to adsorb AV7 and BG, respectively. The Si-O-Si bend lying at 478 cm^{-1} of the prepared RHA–CFA adsorbent shift to the low wavenumber 470 cm^{-1} indicated the involvement of functional group in the adsorption process. The peak at 790 cm^{-1} may be $Si(CH_3)_3$ with CH_3 rock or may also show the presence of primary amine saturation or aromatic NH_2 for spent adsorbents [38].

The peak that lies at 1620 cm^{-1} shows the presence of carbonyl group (C=O), while that at wavenumber 1597 cm^{-1} indicates carboxylate asymmetric -CO_2 stretch. The lack of shift at 1384 cm^{-1} shows that the split CH_3 umbrella mode is not involved in the adsorption process. The peak at 1384 cm^{-1} also indicates the presence of primary amides and signifies C-N stretching and N-H in-plane bending possibly due to the -NO_2 group symmetric stretching. Two bands at wavenumber 1354 and 621 cm^{-1} suggest the presence of compound C-OH, where the peaks shift to 1357 and 617 cm^{-1}, respectively, for AV7 dye adsorption. These peaks do not shift for BG dye adsorption, suggesting the absence of reaction at this outer functional group [25, 38]

X-Ray Diffraction (XRD)

The XRD patterns of the prepared and spent RHA–CFA adsorbents were presented in Fig. (**10**). Almost similar XRD patterns were observed between all the adsorbents. The peaks indicate the presence of crystalline substances, while each substance provided different diffraction patterns [39]. The results show that the prepared RHA–CFA adsorbents comprised cristobalite, quartz, and mullite. Slightly changes in the peaks were observed after the adsorption of AV7 and BG dyes. The peak intensity (not shown in the figure) of spent RHA–CFA adsorbent was increased at $2\theta = 21.5°$, $26°–27°$, and $37.5°$ for adsorption of AV7 while decreased for BG adsorption at $2\theta = 27°$ and $37.5°$, suggesting changes in the bulk phase [40]. Both spent adsorbent peaks shifted slightly to the left after dye adsorption as shown in Fig. (**10**).

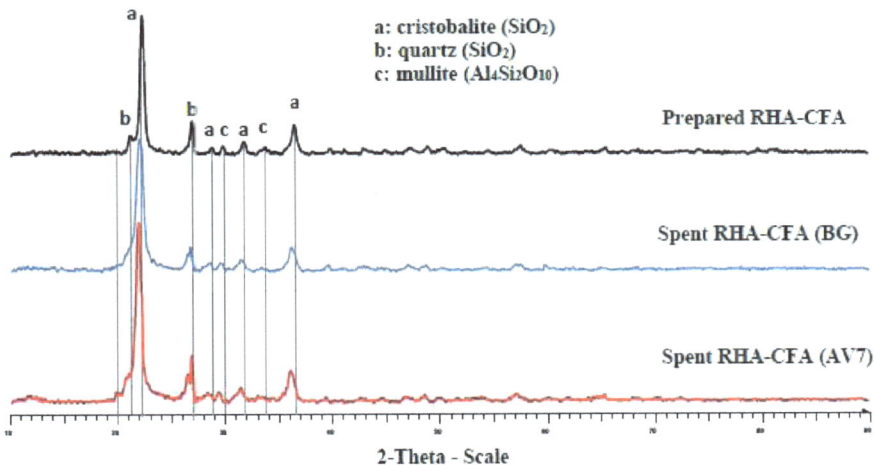

Fig. (**10**). XRD pattern for prepared and spent RHA-CFA adsorbents.

CONCLUDING REMARKS

In the present study, the effectiveness of RHA–CFA adsorbents for AV7 and BG dyes in aqueous media was investigated as function of adsorbent dosage, initial concentration of synthetic dyes, contact time, pH of the solution, shaking speed, and temperature. Comparative results demonstrated that the dosage of adsorbent, initial concentration of synthetic dyes, contact time, pH of the solution, shaking speed, and temperature have important influences on the adsorption of AV7 and BG dyes. XRF, SEM, FTIR, and XRD techniques were used to characterize the prepared and spent RHA–CFA adsorbent. The present work denoted that RHA–CFA has remarkable potential for the removal of AV7 and BG dyes from polluted water.

CONSENT FOR PUBLICATION

Not applicable.

CONFLICT OF INTEREST

The author declares no conflict of interest, financial or otherwise.

ACKNOWLEDGEMENTS

The authors gratefully acknowledge the financial support from the Ministry of Higher Education Malaysia for Fundamental Research Grant Scheme (FRGS) with Project Code: FRGS/1/2019/TK02/USM/02/2 and support received from Universiti Sains Malaysia.

REFERENCES

[1]   A.A. Adeyemo, I.O. Adeoye, and O.S. Bello, "Adsorption of dyes using different types of clay: a review", *Appl. Water Sci.,* vol. 7, no. 2, pp. 543-568, 2017.
[http://dx.doi.org/10.1007/s13201-015-0322-y]

[2]   A. Mary Ealias, and M.P. Saravanakumar, "A critical review on ultrasonic-assisted dye adsorption: Mass transfer, half-life and half-capacity concentration approach with future industrial perspectives", *Crit. Rev. Environ. Sci. Technol.,* vol. 49, no. 21, pp. 1959-2015, 2019.
[http://dx.doi.org/10.1080/10643389.2019.1601488]

[3]   P. Sharma, H. Kaur, M. Sharma, and V. Sahore, "A review on applicability of naturally available adsorbents for the removal of hazardous dyes from aqueous waste", *Environ. Monit. Assess.,* vol. 183, no. 1-4, pp. 151-195, 2011.
[http://dx.doi.org/10.1007/s10661-011-1914-0] [PMID: 21387170]

[4]   V. Katheresan, J. Kansedo, and S.Y. Lau, "Efficiency of various recent wastewater dye removal methods: A review", *J. Environ. Chem. Eng.,* vol. 6, no. 4, pp. 4676-4697, 2018.
[http://dx.doi.org/10.1016/j.jece.2018.06.060]

[5]   D.A. González-Casamachin, J.R. De la Rosa, C.J. Lucio-Ortiz, D.A.D.H. De Rio, and D.X. Martínez-Vargas, "G.A. Flores-Escamilla N.E.D Guzman, V.M. Ovando-Medina, and E. Moctezuma-Velazquez, "Visible-light photocatalytic degradation of acid violet 7 dye in a continuous annular reactor using ZnO/PPyphotocatalyst: Synthesis, characterization, mass transfer effect evaluation and kinetic analysis"", *Chem. Eng. J.,* vol. 373, pp. 325-337, 2019.
[http://dx.doi.org/10.1016/j.cej.2019.05.032]

[6]   H. Ben Mansour, R. Mosrati, D. Corroler, K. Ghedira, D. Barillier, and L. Chekir, "In vitro mutagenicity of Acid Violet 7 and its degradation products by Pseudomonas putida mt-2: Correlation with chemical structures", *Environ. Toxicol. Pharmacol.,* vol. 27, no. 2, pp. 231-236, 2009.
[http://dx.doi.org/10.1016/j.etap.2008.10.008] [PMID: 21783945]

[7]   D. Fabbri, P. Calza, and A.B. Prevot, "Photoinduced transformations of acid violet 7 and acid green 25 in the presence of TiO2 suspension", *J. Photochem. Photobiol. Chem.,* vol. 213, no. 1, pp. 14-22, 2010.
[http://dx.doi.org/10.1016/j.jphotochem.2010.04.014]

[8]   H. Ben Mansour, Y. Ayed-Ajmi, R. Mosrati, D. Corroler, K. Ghedira, D. Barillier, and L. Chekir-Ghedira, "Acid violet 7 and its biodegradation products induce chromosome aberrations, lipid peroxidation, and cholinesterase inhibition in mouse bone marrow", *Environ. Sci. Pollut. Res. Int.,* vol. 17, no. 7, pp. 1371-1378, 2010.
[http://dx.doi.org/10.1007/s11356-010-0323-1] [PMID: 20369386]

[9]   B.H. Hameed, "Equilibrium and kinetic studies of methyl violet sorption by agricultural waste", *J. Hazard. Mater.,* vol. 154, no. 1-3, pp. 204-212, 2008.
[http://dx.doi.org/10.1016/j.jhazmat.2007.10.010] [PMID: 18023971]

[10]   A. Mittal, D. Kaur, and J. Mittal, "Applicability of waste materials—bottom ash and deoiled soya—as adsorbents for the removal and recovery of a hazardous dye, brilliant green", *J. Colloid Interface Sci.,* vol. 326, no. 1, pp. 8-17, 2008.
[http://dx.doi.org/10.1016/j.jcis.2008.07.005] [PMID: 18675425]

[11]   P. Sharma, R. Kaur, C. Baskar, and W.J. Chung, "Removal of methylene blue from aqueous waste using rice husk and rice husk ash", *Desalination,* vol. 259, no. 1-3, pp. 249-257, 2010.
[http://dx.doi.org/10.1016/j.desal.2010.03.044]

[12]   K.D. Rao, P.R.T. Pranav, and M. Anusha, "Stabilization of expansive soil with rice husk ash, lime and gypsum – an experimental study", *Int. J. Eng. Sci. Technol.,* vol. 3, no. 11, pp. 8076-8086, 2012.

[13]   FAO, *Crop 2018 statistics.*http://www.fao.org/faostat/en/#data/QC.

[14]   R. Pode, "Potential applications of rice husk ash waste from rice husk biomass power plant", *Renew. Sustain. Energy Rev.,* vol. 53, pp. 1468-1485, 2016.
[http://dx.doi.org/10.1016/j.rser.2015.09.051]

[15]   M. Rafieizonooz, J. Mirza, M.R. Salim, M.W. Hussin, and E. Khankhaje, "Investigation of coal bottom ash and fly ash in concrete as replacement for sand and cement", *Constr. Build. Mater.,* vol. 116, pp. 15-24, 2016.
[http://dx.doi.org/10.1016/j.conbuildmat.2016.04.080]

[16]   I. Dahlan, and S. Mahdzir, "Adsorption of dyes in aqueous medium using RHA and CFA: Effect of preparation methods and process optimization", In: *Handbook of Research on Resource Management for Pollution and Waste Treatment.,* A.C. Affam, E.H. Ezechi, Eds., IGI Global: Pennsylvania, USA, 2019, pp. 458-475.

[17]   I. Dahlan, and N.W. Ling, "Adsorption of acid violet 7 (AV7) dye using RHA-CFA adsorbent", *Modelling, process analysis and optimization", Sep. Sci. Technol.,* Latest articles 2019.
[http://dx.doi.org/10.1080/01496395.2019.1708115]

[18]   L. Ai, H. Huang, Z. Chen, X. Wei, and J. Jiang, "Activated carbon/CoFe2O4 composites: Facile synthesis, magnetic performance and their potential application for the removal of malachite green from water", *Chem. Eng. J.,* vol. 156, no. 2, pp. 243-249, 2010.
[http://dx.doi.org/10.1016/j.cej.2009.08.028]

[19]   B.K. Nandi, A. Goswami, and M.K. Purkait, "Adsorption characteristics of brilliant green dye on kaolin", *J. Hazard. Mater.,* vol. 161, no. 1, pp. 387-395, 2009.
[http://dx.doi.org/10.1016/j.jhazmat.2008.03.110] [PMID: 18456401]

[20]   S. Chowdhury, R. Mishra, P. Saha, and P. Kushwaha, "Adsorption thermodynamics, kinetics and

isoteric heat of adsorption of malachite green onto chemically modified rice husk", *Desalination,* vol. 265, no. 1-3, pp. 159-168, 2011.
[http://dx.doi.org/10.1016/j.desal.2010.07.047]

[21]   H. Ben Mansour, D. Corroler, D. Barillier, K. Ghedira, L. Chekir, and R. Mosrati, "Evaluation of genotoxicity and pro-oxidant effect of the azo dyes: Acids yellow 17, violet 7 and orange 52, and of their degradation products by Pseudomonas putida mt-2", *Food Chem. Toxicol.,* vol. 45, no. 9, pp. 1670-1677, 2007.
[http://dx.doi.org/10.1016/j.fct.2007.02.033] [PMID: 17434654]

[22]   V.S. Mane, I. Deo Mall, and V. Chandra Srivastava, "Kinetic and equilibrium isotherm studies for the adsorptive removal of Brilliant Green dye from aqueous solution by rice husk ash", *J. Environ. Manage.,* vol. 84, no. 4, pp. 390-400, 2007.
[http://dx.doi.org/10.1016/j.jenvman.2006.06.024] [PMID: 17000044]

[23]   D.O. Cooney, *Adsorption Design for Wastewater Treatment.* Lewis Publishers, CRC Press LLC: Boca Raton, USA, 1999.

[24]   I.D. Mall, V.C. Srivastava, and N.K. Agarwal, "Adsorptive removal of Auramine-O: Kinetic and equilibrium study", *J. Hazard. Mater.,* vol. 143, no. 1-2, pp. 386-395, 2007.
[http://dx.doi.org/10.1016/j.jhazmat.2006.09.059] [PMID: 17074434]

[25]   U.R. Lakshmi, V.C. Srivastava, I.D. Mall, and D.H. Lataye, "Rice husk ash as an effective adsorbent: Evaluation of adsorptive characteristics for Indigo Carmine dye", *J. Environ. Manage.,* vol. 90, no. 2, pp. 710-720, 2009.
[http://dx.doi.org/10.1016/j.jenvman.2008.01.002] [PMID: 18289771]

[26]   L. Borah, M. Goswami, and P. Phukan, "Adsorption of methylene blue and eosin yellow using porous carbon prepared from tea waste: Adsorption equilibrium, kinetics and thermodynamics study", *J. Environ. Chem. Eng.,* vol. 3, no. 2, pp. 1018-1028, 2015.
[http://dx.doi.org/10.1016/j.jece.2015.02.013]

[27]   K.M. Sreenivas, M.B. Inarkar, S.V. Gokhale, and S.S. Lele, "Re-utilization of ash gourd (Benincasa hispida) peel waste for chromium (VI) biosorption: Equilibrium and column studies", *J. Environ. Chem. Eng.,* vol. 2, no. 1, pp. 455-462, 2014.
[http://dx.doi.org/10.1016/j.jece.2014.01.017]

[28]   N. Rajamohan, M. Rajasimman, R. Rajeshkannan, and V. Saravanan, "Equilibrium, kinetic and thermodynamic studies on the removal of Aluminum by modified Eucalyptus camaldulensis barks", *Alex. Eng. J.,* vol. 53, no. 2, pp. 409-415, 2014.
[http://dx.doi.org/10.1016/j.aej.2014.01.007]

[29]   N.A. Kabbashi, M.A. Atieh, A. Al-Mamun, M.E.S. Mirghami, M.D.Z. Alam, and N. Yahya, "Kinetic adsorption of application of carbon nanotubes for Pb(II) removal from aqueous solution", *J. Environ. Sci. (China),* vol. 21, no. 4, pp. 539-544, 2009.
[http://dx.doi.org/10.1016/S1001-0742(08)62305-0] [PMID: 19634432]

[30]   K.P. Singh, D. Mohan, S. Sinha, G.S. Tondon, and D. Gosh, "Color removal from wastewater using low-cost activated carbon derived from agricultural waste material", *Ind. Eng. Chem. Res.,* vol. 42, no. 9, pp. 1965-1976, 2003.
[http://dx.doi.org/10.1021/ie020800d]

[31]   S. Banerjee, and M.C. Chattopadhyaya, "Adsorption characteristics for the removal of a toxic dye, tartrazine from aqueous solutions by a low cost agricultural by-product", *Arab. J. Chem.,* vol. 10, pp. S1629-S1638, 2017.
[http://dx.doi.org/10.1016/j.arabjc.2013.06.005]

[32]   V. Ponnusami, V. Krithika, R. Madhuram, and S.N. Srivastava, "Biosorption of reactive dye using acid-treated rice husk: Factorial design analysis", *J. Hazard. Mater.,* vol. 142, no. 1-2, pp. 397-403, 2007.
[http://dx.doi.org/10.1016/j.jhazmat.2006.08.040] [PMID: 17011118]

[33]   J.R. Baseri, P.N. Palanisamy, and P. Sivakumar, "Comparative studies of the adsorption of direct dye on activated carbon and conducting polymer composite", *J. Chem.,* vol. 9, no. 3, pp. 1122-1134, 2012.

[34]   Y. Bulut, N. Gözübenli, and H. Aydın, "Equilibrium and kinetics studies for adsorption of direct blue 71 from aqueous solution by wheat shells", *J. Hazard. Mater.,* vol. 144, no. 1-2, pp. 300-306, 2007.
[http://dx.doi.org/10.1016/j.jhazmat.2006.10.027] [PMID: 17118540]

[35]   K.N. Farooque, M. Zaman, E. Halim, S. Islam, M. Hossain, Y.A. Mollah, and A.J. Mahmood, "Characterization and utilization of rice husk ash (RHA) from rice mill of Bangladesh", *Bangladesh J. Sci. Ind. Res.,* vol. 44, no. 2, pp. 157-162, 1970.
[http://dx.doi.org/10.3329/bjsir.v44i2.3666]

[36]   H. Misran, R. Singh, S. Begum, and M.A. Yarmo, "Processing of mesoporous silica materials (MCM-41) from coal fly ash", *J. Mater. Process. Technol.,* vol. 186, no. 1-3, pp. 8-13, 2007.
[http://dx.doi.org/10.1016/j.jmatprotec.2006.10.032]

[37]   G. Bai, Y. Qiao, B. Shen, and S. Chen, "Thermal decomposition of coal fly ash by concentrated sulfuric acid and alumina extraction process based on it", *Fuel Process. Technol.,* vol. 92, no. 6, pp. 1213-1219, 2011.
[http://dx.doi.org/10.1016/j.fuproc.2011.01.017]

[38]   B.C. Smith, *Infrared Spectral Interpretation: A Systematic Approach.* CRC Press LLC: Florida, USA, 1998.

[39]   "Scintag Inc", *Chapter 7: Basics of X-Ray Diffraction.*http://www.geo.umass.edu/courses/geo311/xrdbasics.pdf

[40]   P. Sathishkumar, M. Arulkumar, and T. Palvannan, "Utilization of agro-industrial waste Jatropha curcas pods as an activated carbon for the adsorption of reactive dye Remazol Brilliant Blue R (RBBR)", *J. Clean. Prod.,* vol. 22, no. 1, pp. 67-75, 2012.
[http://dx.doi.org/10.1016/j.jclepro.2011.09.017]

CHAPTER 8

Pollution Prevention Assessments: Approaches and Case Histories

Ashok Kumar[1,*], **Saisantosh Vamshi Harsha Madiraju**[1] and **Lakshika Nishadhi Kuruppuarachchi**[1]

[1] *The College of Engineering , The University of Toledo, Toledo, OH, USA43606*

Abstract: The pollution prevention (P2) approach known as source reduction is being used worldwide to reduce the deleterious effects on human health and the environment due to the contaminants released from a variety of industrial sources. This chapter focuses on the concept of pollution prevention approaches undertaken by the U.S.EPA. P2 approach is discussed by applying the concept of energy efficiency, energy savings, greenhouse gas emission (GHG) reductions, waste reduction, and stormwater management to local schools, restaurants, hospitals, and the industrial sector in Ohio, USA. Several publicly available tools were used to analyze data collected during assessments. The major tools used are the Energy Assessment Spreadsheet tool (developed by Air Pollution Research Group at the College of Engineering, The University of Toledo, Ohio, USA) for the energy savings and Economic Input Life Cycle Assessment tool (developed by researchers at the Green Design Institute of Carnegie Mellon University) for the estimation of environmental emissions from industrial activities.These approaches result in the reduction of financial costs for waste management, cleanup, health problems, and environmental damage. Outcomes of pollution prevention activities are knowledge-based, behavioral, health-related, or environmental, which includes decreased exposure to toxins, conservation of natural resources, decreased release of toxins to the environment, and cost savings. The chapter presents case studies that focused on energy, greywater reuse, and food waste diversion from landfills.

Keywords: Approaches, Assessment, Pollution prevention, Case studies, Cost savings, Energy savings, Industrial facilities, USEPA, Wastewater.

INTRODUCTION

The concept of reducing or eliminating pollution at the source is called pollution prevention also known as P2. Considering the impact on the environment along with the amount of money, time, and resources involved in the disposal of waste generated, the USEPA encouraged the industry to apply the P2 approach to reduce

* **Corresponding author Ashok Kumar:** The College of Engineering , The University of Toledo, Toledo, OH, USA, 43606; Tel: 419-934-0878; E-mail: akumar@utoledo.edu

G. Venkatesan, S. Lakshmana Prabu and M. Rengasamy (Eds.)

the release of contaminants to air, water, and land media [1]. The amount of waste to be treated, controlled, and disposed of is less when there is a decrease in the amount of pollution produced. The reduced pollution, which is covered in pollution prevention activities, also implies a low risk to public health as well as the environment. Pollution prevention activities were carried out in these industries like reducing pollution at the source by modification of manufacturing processes, switching to eco-friendly maintenance use chemicals, following conservation practices [2]. All kinds of pollution-generating activities can implement these pollution prevention approaches, which include the energy, agriculture, consumer, and industrial sectors. Pollution prevention reduces both financial costs, including waste management & cleanup, and environmental costs and costs associated with health problems & environmental damage [3]. One of the main advantages of P2 is to protect and conserve the environment by optimal use of natural resources. They also strengthen economic growth by improving ways for efficient production following eco-friendly alternatives in the industrialfacilities [4]. These activities help households, businesses, and communities to handle the waste. Knowledge-based, behavioral, health-related, or environmental targets are the outcomes of pollution prevention activities. They have decreased exposure to toxins, conservation of natural resources, decreased release of toxins to the environment, energy, and cost savings [5].

The way each industrial facility operates varies with the type of facility they fall under. Examples of different types of facilities are food processing, manufacturing, automobile, furniture, chemical, universities, laboratories, *etc.* The amount and type of waste generated and pollution created also vary. But the P2 approach stands first in the place where reduction of pollution at the source. This approach is the most preferable and most efficient way to reduce pollution [6]. (Fig. **1**) represents the waste management structure for pollution generation activities that need to be preferred in the industrial, agricultural, energy, consumer, and federal sectors to control pollution, according to the Encyclopedia of Chemical Processing [7].

The USEPA is creating awareness by educating the people through its resources about the importance of P2, laws, and policies created, actions taken by EPA through websites, conferences, webinars, training workshops, *etc.* The EPA funds different programs (*e.g.*: Pollution Prevention Grant Program and Source Reduction Assistance Grant Program (SRA)) [8, 9] to assist the local industrial facilities freely and implement pollution prevention approaches. National Emphasis areas (NEAs) are program priorities for the P2 program.

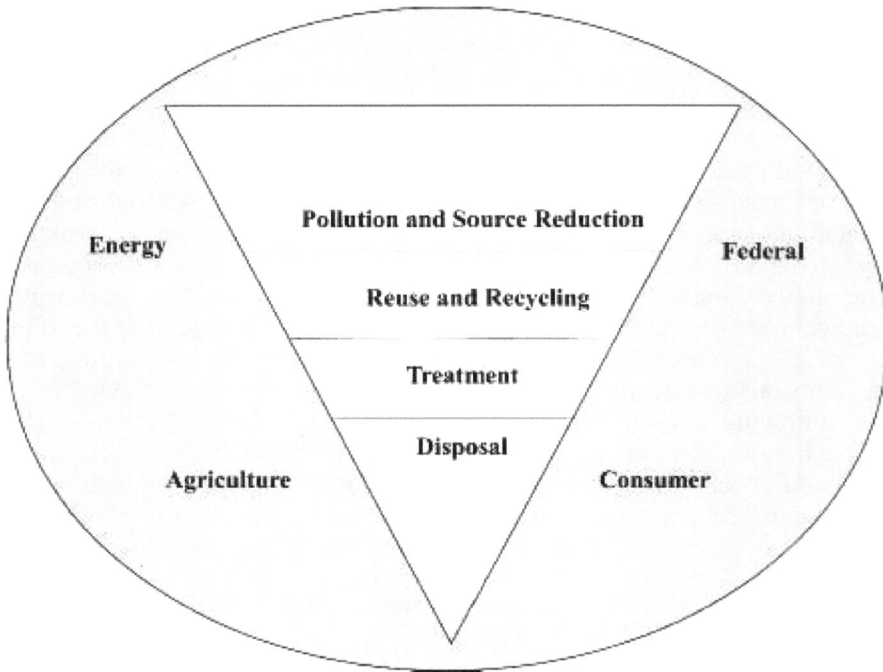

Fig. (1). The waste management structure for pollution generation activities.

POLLUTION PREVENTION

What is P2?

According to the Pollution Prevention Act of 1990, P2 is defined as "Any practice that reduces the amount of any hazardous/toxic substance, pollutant, or contaminant entering any waste stream or otherwise released into the environment including fugitive emissions before recycling of discarded material, treatment, or disposal; and reduces the hazards to public health and the environment associated with the releases of those substances, pollutants or contaminants" [7]. P2 approach comprises practices that increase the efficient use of water, energy use, or use of raw materials, or taking other actions that protect natural resources before recycling, clean up or disposal and practices that may protect natural resources through conservation methods, or in-process recycling (*i.e.*, process improvements to reuse materials within the same business/ facility in the production process) [9].

What is not P2?

P2 practices alter the characteristics and quantity of the pollutant or a toxic substance through a process that itself is not integral to providing a service or necessary to produce a product. Activities like recycling and disposal activities, cleaning up waste, non-hazardous solid waste management practices are not P2 activities [7, 9]. Examples of non-P2 practices include recycling plastic, office paper, glass, cardboard, and metal, *etc.*

P2. Practice Examples

Some examples are:

- Reduce air emissions - steps to prevent ammonia refrigeration leaks and other fugitive releases, substitute the use of aqueous material for volatile materials, byproduct recovery, and reused within that business manufacturing processes [10].
- Reduce the number of pollutants discharged from chemical and manufacturing facilities to water bodies and recover organic and inorganic materials to prevent water contamination while cleaning [11, 12].
- Reduce energy use and increase energy efficiency - heat transfer systems, equipment upgrades/innovations, process changes/innovations, cleaner fuels [12].
- Conserve water - use high-volume, low-pressure washing systems that reuse water; identify practices to minimize loadings to wastewater systems, extend production line times between cleanings to minimize water use [13].
- Use low-toxicity products-for ingredients, refrigerants, building cleaners, food-grade lubrication, and processing techniques and equipment [14].

Pollution Prevention Approaches

The P2 projects in any area are implemented to aid small and mid-size manufacturers and other entities through training in source reduction techniques and direct provision of pollution prevention technical assistance including environmental management systems, clean manufacturing, and energy efficiency assessments [15]. The goal is to help the industry to reduce pollution using available pollution prevention tools.

The contribution made by the agencies and researchers before and after passing the Pollution Prevention Act in 1990 helped the public to create awareness about the implementation of pollution prevention activities [16]. Some potential reports in the past three decades are identified and listed in Fig. (**5**). These reports mainly contain the guides for planning, operations, implementation, using P2 tools, P2 concepts and practices, management measures, pollution solutions, re-evaluating existing P2 programs, and guide sheets [17 - 30]. The report developed from the National Environmental Justice Advisory Council (NEJAC) Meeting, conventional pollution prevention contains a wide variety of activities Fig. (**2**) [31]. The facility pollution prevention guide 1992 [18] stated that an effective pollution prevention program will address the conservation of the environment, public health, and industrial development (Fig. **3**) According to the "FY 2020 and FY 2021 Pollution Prevention Grant Program" by USEPA, outcome measures used in current P2 programs are shown in Fig. (**4**) [32].

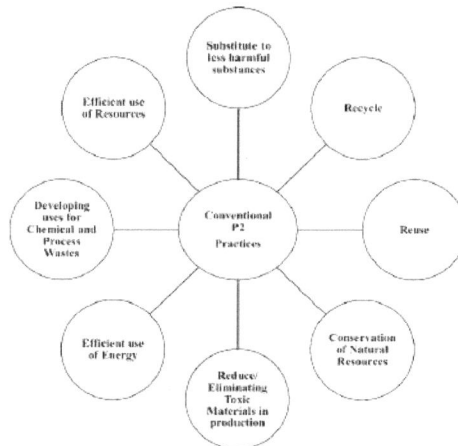

Fig. (2). Conventional pollution prevention practices according to NEJAC.

Fig. (3). The focus of effective pollution prevention program according to facility pollution prevention guide 1992.

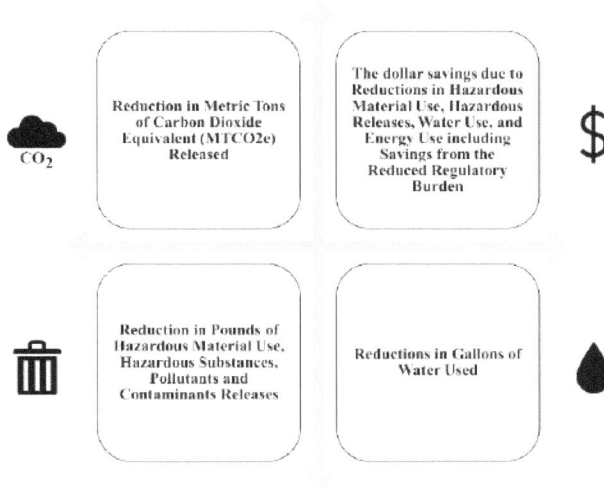

Fig. (4). Outcomes of FY 2020 and FY 2021 pollution prevention grant program.

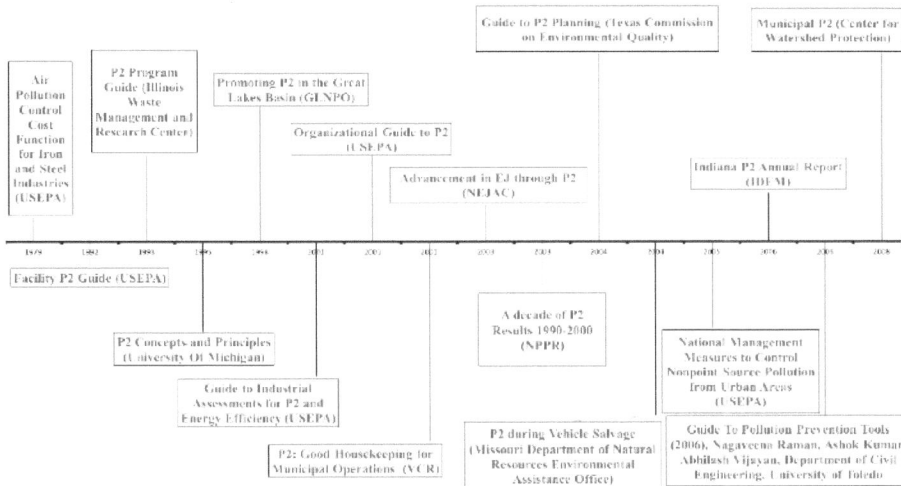

Fig. (5). Timeline of pollution prevention research reports in the past three decades.

Pollution Prevention is an important voluntary tool to be used by the facilities to reduce their expenditure on the pollution/waste created and the same disposal [33]. These guides provide a strong inspiration to the facilities/businesses to initiate pollution prevention approaches. Success stories of the facilities that implemented P2 approaches are shared on the EPA P2 website [34] and the people who participated have published peer-reviewed publications, which is more likely to create awareness for the small businesses, and upcoming employees to aspire to pollution prevention solutions.

To measure the results/outcomes of P2 activities, including both economic and environmental, spreadsheets are designed and made available online by USEPA [35, 36]. That include:

- P2 Cost Calculator
- Gallons to Pounds Converter
- P2 Greenhouse Gas Calculator

These spreadsheets are publicly available and can be utilized by the people who are participating in the P2 activities. Using the P2 Cost Calculator tool, reductions in cost for hazardous wastes, air permitting fees, water, and electricity charges. The Gallons to Pounds Converter is used to convert different common business units into P2 grant reporting units. P2 GHG calculator is used to Calculate GHG emission reductions from conservation of electricity, water, fuel, and chemical substitutions, green energy, and improved materials management. To draw on data in the P2 Cost Savings Calculator, the GHG calculator has a feature that makes a direct conversion of GHG reductions to related cost savings [36]. Additional tools were also made available which are included under products and services [35], which are:

- EPA'S ENERGY STAR portfolio manager
- Waste Reduction Model (WARM)
- Electronics Environmental Benefits Calculator (EEBC)
- Green Cleaning Pollution Prevention Calculator
- Hospitality Environmental Benefits Calculator

The Air Pollution Research Group at The University of Toledo developed pollution prevention tools under the guidance of Ohio Statewide Environmental Network [37], which include:

- GAP (1.0)
- MSM (1.1 & 1.2)
- ERC (1.0)
- LEAN (1.0
- HVAC Checklist (1.0)
- Energy Assessment Spreadsheet (1.0)
- Hybrid HVAC System Design Tool (Ver.1.0)
- Building Sustainability Tool (Version 1.0)
- Hospital Assessment Tool (Version 1.0)
- Database for Green Products (DGP Version 1.0)
- Small Business Self-Assessment Tool (SBSAT Version 1.0)

- Department Specific Hospital Assessment Tool (D-HAT Version 1.0)
- Chemical Identification Software (CIS Version 1.0)
- Food Assessment Tool (FAT Version 1.0)
- Sustainable Hospital Assessment Tool (S-HAT Version 1.0)

Case History of Assessments

This section provides the general framework and overview of the P2 assessments performed for small and mid-size manufacturers and other entities. The combined assessments performed for pollution prevention and energy-saving were described to guide those performing P2 assessments at any facility with the source of pollution. The assessments were conducted at some voluntarily supported facilities that include hospitals, automobile assembly plants, schools, hardware workshops, correctional institutes, restaurants, and food processing facilities in Ohio. Based on the walk-through surveys and data collection from each facility, the appropriate P2 approaches were determined for each of them. For the energy efficiency and energy-saving assessments, a survey was carried out at the facilities by recording the information about the existing lighting fixtures and electric equipment. The stepwise procedure is briefly shown in Fig. **(6).** The EAS was then used to estimate the usage cost per year for the existing and the recommended fixtures/equipment. After that, an LCA was carried out using the EIO-LCA tool on the existing and recommended fixtures/equipment for comparison [38]. The estimated and anticipated savings from The University of Toledo P2 assessments are mentioned in Table **1**. The difference in estimated and anticipated readings is mainly due to the delay in implementation by the industries. The first potential reason for the delay is the budget year. If the assessments are planned at the end of the budget year. The implementation is made in the next budget year due to funds management. The second potential reason is if alternative materials are suggested for the replacement of current materials, industries wait for their current purchased stock of materials to complete for the replacement with alternative materials. The third potential reason is to calculate the annual savings, need to wait at least one year from the date of implementation to see the anticipated saving.

Table 1. ACS Metrics of estimate and anticipated reductions due to P2 assessments by The University of Toledo.

ACS Metric	Estimated Reductions	Anticipated Reductions
Hazardous Material/ Waste Reduced (lbs)	40,670	20,235
Water Saved (gallons)	722,700	120,000

(Table 1) cont.....

ACS Metric	Estimated Reductions	Anticipated Reductions
GHG Reduced (MTCO2e)	505	143
Cost Savings ($)	660,000	180,000
Electricity Saved-(kwh)	6,120	2,000

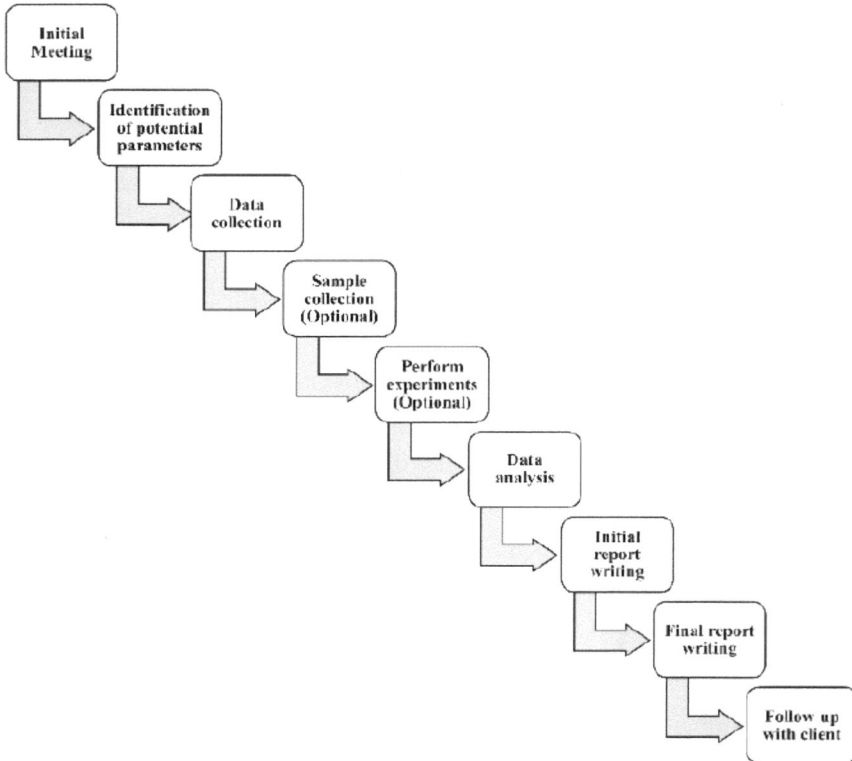

Fig. (6). The step-by-step procedure was followed in completing the P2 assessments.

This section discusses some of the objectives and P2 recommendations made to the facilities in Ohio, covering topics on energy, and cost savings, water savings, hazardous material/waste reduction, Green House Gas savings.

The objective of the assessments conducted at the facilities was determined after the initial site visit. Then potential parameters to imply the P2 approach were identified. In-site data required to perform the assessments were collected, previous records of purchase orders, electricity bills, and necessary files were requested. Records of purchase orders is a piece of helpful information to get an idea about the current financial ability spent for suggesting better affordable buying options on replacements. Written reports with assessment findings and recommendations were submitted. Also, any supporting analysis including, spreadsheets and documents, were submitted to the facility. Additional follow-up and information exchange are continued to record the anticipated reductions and savings were recorded.

Energy and Cost Savings at a Manufacturing Facility

The objective of the assessment is to provide improvements in lighting and electricity usage for energy and cost savings.

The following recommendations were made with estimated reductions mentioned in Table **2**.

- Existing T8 fluorescent fixtures with ballast may not be compatible with all replacement LED tubes (requires fluorescent ballast disconnect/bypass and maybe end receptacle replacement).
- If installation and labor costs can be kept minimal, the return on investment for $20 LED grow-lights (4-feet 18-watt tubes) is about 2 years.
- It is recommended that a set of 2 to 4 LED grow light tubes could be installed for a trial period of 3-6 months and if the performance is found satisfactory, a total 18 tubes could be replaced.

Table 2. Comparison of existing and recommended lighting fixture annual costs.

Fixture Type	Existing	Recommended
Wattage	40	18
Quantity	220	220
Total Wattage	8800	3960
Total Hours / Year	2080	2080
kWh / Year	18304	8237
Cost / Year	1830.4	823.68
Total kW / Year	9	4
Total kWh / Year	18304	8237
Total Cost / Year ($)	1830.4	823.68

Cost of Energy Considered in Existing and Recommended Fixtures = 0.10$/kWh.

This assessment was conducted and recommended based on similar studies performed and implemented in real-time. A similar analysis was conducted by Michael Schratz in 2013 to identify the benefits of shifting from conventional lighting to LEDs in manufacturing facilities. His study also compared LED lighting with the other existing lighting technologies and found that the LED creates less risk of fire, product contamination, energy savings, durability with a better performance. After installing LED lighting in 10 facilities, their investment is paid back in approximately 29 months with an annual savings of $6,491,259 [39].

Water savings at a University

The objective is to propose a greywater treatment system to the UT Engineering buildings and evaluate project implementation cost, reductions in water consumption, and cost reductions. Wastewater collected from sinks and washing equipment that would otherwise be sent down the drain and into the sewage system. Grey water is not potable water, but it can be used for some activities such as flushing toilets and gardening [40]. The reason behind the proposal of this issue is to re-use greywater to save freshwater usage that comes underwater conservation practices. It is a well-known fact that only 3% of the surface water is freshwater and only 0.5% of surface water and approximately 0.3% of groundwater is available for drinking [41]. In this emerging world of overpopulation, water is an essential natural resource to be conserved [42]. This objective is under the activities.

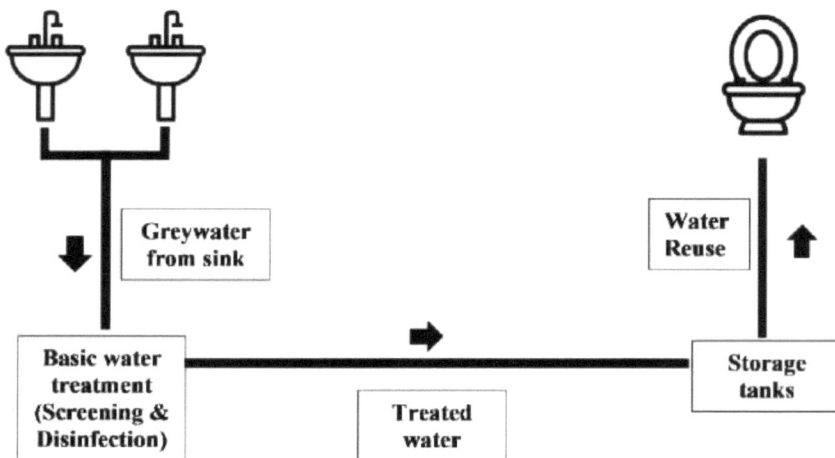

Fig. (7). The layout of the proposed model for greywater reuse.

The proposed model is the most efficient model. The process starts with consideration of one washroom for the greywater treatment and reuse (see Fig. 7). The average greywater from the multiple sinks in a washroom is calculated. The basic water treatment system is designed based on the average daily grey water capacity. The treatment system involves screening and disinfection. The treated water is stored in storage tanks designed according to the 1.5 X maximum capacitydesign [43, 44]. The water is re-used in different ways [45]:

- In the toilet flush
- Water in the outdoor garden plants on campus
- In growing indoor plants

This can be useful in reducing water/wastewater bills along with minimizing water pollution. The total estimated amount of water to be treated and re-used per year is 722,700 gallons/ year.

E. Friedler conducted a study on the on-site greywater treatment and its reuse in multi-story buildings in 2005. His study provided evidence that a combination of low organic matter, microbial indicators, and nutrients minimizes fouling potentials, regrowth and ensures treated greywater can be reused safely for toilet flushing [46]. A similar experimental investigation was conducted by A. Campisano and C. Modica in 2010 to reuse greywater from the washbasin for toilet flushing. This study provides real-time evidence that wastewater reduction and freshwater savings can be achieved by reusing greywater. The results of the study indicate that if the greywater storage tank is five times greater than the toilet cistern, then a minimum of 32% of water savings can be achieved considering the greywater detention time [47]. As per the literature, this assessment was designed and recommended.

Other Environmental Benefits of the Proposed Model

- Reduce the freshwater extractions from surface water sources.
- Increasing the fertility of the topsoil in the garden.
- Reduce the burden on community wastewater plants.
- Groundwater replenishment.
- Can overcome some dry soil problems.

Waste and GHG Emission Reduction at a Hospital

A hospital that provides health services is open 24 hours/day. The estimated number of meals served in the cafeteria area is 1400/day. The current practice is to use Styrofoam boxes and plates when serving food. Food soiled Styrofoam is not recyclable, and it's sent to landfill. As an alternative, bio-degradable lunch boxes are recommended, and these can be sent to a composting facility [48]. The waste generated from Styrofoam usage will be reduced.

Food waste in the cafeteria area and the kitchen area is currently sent to a landfill. Sending the food waste to a local composting facility is recommended and this will reduce the food waste going to landfills and eventually reduce GHG emission [49].

There are two main objectives of this assessment:

1. To minimize the utilization of Styrofoam plates and boxes at Mercy Hospital food court by replacing them with bio-degradable materials.

2. Food waste diversion from landfill to a composting facility and reduce GHG emission.

Total estimated food waste per year = (0.6*511,000) *0.0005 = 153 ton/year.

Net GHG emission saving from diverting food waste from combustion to composting= 0.18*153 MtCo2e /year= 27.54 MtCo2e/year.

This assessment is based on the corresponding literature. A research study provided by Tim A. (2015) in Food Production Systems & Sustainability shows the comparison between landfills and composting by estimating the GHG emissions and other environmental effects. The study states that composting food waste may be the most effective way to drive change and it is the better environmentally sustainable food waste management approach [50]. A study on biodegradable alternatives for styrofoam and plastics at "St. Jerome's" in 2003 shows the necessity to replace styrofoam with an environmentally friendly biodegradable alternative to minimize soil contamination [51]. A study on innovative technologies for biodegradable packaging at Sa Jose State University in 2006 provides a detailed description of bioplastics in food packaging. This study states that the food packaging material made from plant-based resources helps to address environmental issues [52].

EIO-LIFE CYCLE ANALYSIS FOR A CORRECTION INSTITUTE

The objective of this assessment was to evaluate current energy usage and determine additional improvements and GHG emission reduction opportunities by food waste diversion at a correctional institute. Tasks included review of past and current facility utilities records, facility-wide survey and equipment data collection, and interviews with staff, and energy usage modeling and estimation.

The LCA was carried out using the EIO-LCA model for both the current fixtures and recommended Energy Star fixtures [53]. The "power generation and supply" sector was selected in the operational phase to run the EIO-LCA tool. Table **3** provides information regarding the GHGs generated, energy used, and hazardous waste produced for the cases of estimated electricity consumed by the facility before implementing the recommendations and Table **4** provides information regarding the estimated electricity and cost savings after implementing the recommendations.

Table 3. Results of EIO-LCA for both existing and recommended fixtures per year for the Toledo Correction Institute.

Factor (at source)	Existing fixture energy	Energy Star fixture energy	Energy savings ($48,100)
Greenhouse gases (total metric tons of CO_2e)	971	547	424
Energy (terajoules)	11.8	6.67	5.13
Hazardous waste generation (short tons)	1840	1040	800

Cost Savings

It was recommended that the facility likely has to go through some remaining lifespan of the current fluorescent lighting system. However, it is recommended that the facility thoroughly investigate LED lighting in a 2-3-year timeframe. The cost of LED lighting is progressively reducing and financial incentives like grants, *etc.* are becoming available from public and electric utility programs. Even with the low annual operating hours, it could be economical to install a new PE motor instead of rewinding/repairing an AE motor in the event of a breakdown. Because of the low operating hours, the replacement of any existing AE motor with a new PE motor may not be economical.

Table 4. LED lighting and savings estimation.

Fluorescent and LED Lighting Comparison	Estimated Electricity, kWh/Year*	Estimated Cost, $/kWh**
Fluorescent (Existing, 32-Watt 4-Feet Tubes)	1,184,000	$110,100
LED (New Replacement, 18-Watt 4-Feet Tubes)	666,000	$62,000
% Savings	44%	44%
Estimated Replacement Cost ***	$220,000	-
Simple Payback Period, Years	5	-

(Table 4) cont.....

Fluorescent and LED Lighting Comparison	Estimated Electricity, kWh/Year*	Estimated Cost, $/kWh**
* Calculations based on fixtures summary		
** @ 9.3 cents/kWh		
*** Replacement Cost Nominal $80/2-Lamp Fixture (including $20/LED lamp and $40/fixture installation and miscellaneous costs.)		

Estimated cost savings on LED lightening = $110,100 – $62,000 = **$48,100** per year.

SUMMARY

This chapter concludes that P2 is the most preferable approach in the current scenario for industrial facilities and mid-size businesses in terms of pollution prevention and management. In the US, EPA has provided enough facilities, opportunities, and tools for the industrial, agriculture, energy, federal, and consumer sectors to guide, select, follow, implement and recommend P2 approaches and share success for inspiration for upcoming business. This chapter has provided detailed information about the understanding of the P2 approach and what activities fall under the category of the P2 approach. Some of the potential contributions made by the researchers in the past are also discussed. Some of the contributions made by The University of Toledo under the "Pollution Prevention Grant Program" are explained as case histories for a better understanding of the approach.P2 approach was applied on assessments mainly focusing on energy efficiency, energy-saving, GHG emission reduction, and waste reduction. The assessments were done using the EAS, EIO-LCA tools, and other tools developed by the "Air Pollution Research Group" at The University of Toledo under the guidance of Ohio Statewide Environmental Network. Due to the difficulty in accessing the most recent data for some assessments; records of electric equipment, number of employees, records of utility bills, and invoices from waste handlers, were considered from past years for the calculations. Through this chapter, the authors suggest the industries use the cost-free opportunity provided by the U.S. EPA and voluntarily be a part of the pollution prevention program in their states to contribute towards pollution prevention.

CONSENT FOR PUBLICATION

Not applicable.

CONFLICT OF INTEREST

The author declares no conflict of interest, financial or otherwise.

ACKNOWLEDGEMENTS

The pollution Prevention project at the University of Toledo has been funded by the USEPA for carrying out the assessments. We thankthe helpful comments received from Ohio EPA and CIFT during the grant period. Any opinions, findings, and conclusions, or recommendations expressed in this material are those of the authorsand do not necessarily reflect the views of the USEPA, Ohio EPA, or any agency thereof.

REFERENCES

[1]   US EPA, "Pollution Prevention (P2)", *Pollution Prevention (P2)," United States Environmental Protection Agency*. https://www.epa.gov/p2

[2]   J.M. Strock, P.E. Helliker, and D.W. Chan, "Integrated pollution prevention: CAL-EPA's perspective", *Envtl L,* vol. 22, p. 311, 1992.

[3]   T.T. Shen, "Industrial pollution prevention", In: *Industrial Pollution Prevention.* Springer, 1995, pp. 15-35.
[http://dx.doi.org/10.1007/978-3-662-03110-0_2]

[4]   D. Huisingh, and V. Bailey, *Making pollution prevention pay: ecology with economy as policy.* Elsevier, 2013.

[5]   US EPA, "Learn About Pollution Prevention", *United States Environmental Protection Agency*. https://www.epa.gov/p2/learn-about-pollution-prevention (accessed Feb. 03, 2020).

[6]   N. Chadha, "Develop multimedia pollution prevention strategies", *Chem. Eng. Progress United States,* vol. 90, p. 11, 1994.

[7]   S. Lee, *Encyclopedia of chemical processing.* vol. 3. Taylor & Francis US, 2006.

[8]   D. Twickler, "E3: A Growth Strategy For American Manufacturers and Their Communities", *GLRPPR Annual Meeting*, 2012 Chicago. IL 2012.

[9]   US EPA, *Overview of FY20-21 National Emphasis Areas (NEAs) for Pollution Prevention Grants.,* 2020.

[10]   P.M. Randall, "Pollution prevention strategies for the minimizing of industrial wastes in the VCM-PVC industry", *Environ. Prog.,* vol. 13, no. 4, pp. 269-277, 1994.
[http://dx.doi.org/10.1002/ep.670130416]

[11]   S.K. Sharma, and R. Sanghi, *Advances in water treatment and pollution prevention.* Springer Science & Business Media, 2012.
[http://dx.doi.org/10.1007/978-94-007-4204-8]

[12]   V. Moldovan, *Method for developing and promoting operations and services that are supported by an energy, energy efficiency, water management, environmental protection and pollution prevention fund,* 2003.

[13]   S.I. Abou-Elela, F.A. Nasr, H.S. Ibrahim, N.M. Badr, and A.R.M. Askalany, "Pollution prevention pays off in a board paper mill", *J. Clean. Prod.,* vol. 16, no. 3, pp. 330-334, 2008.
[http://dx.doi.org/10.1016/j.jclepro.2006.07.045]

[14]   J.A. Veil, C. Burke. and D. Moses, *Synthetic drilling fluids-a pollution prevention opportunity for the oil and gas industry.* Argonne National Lab: IL, United States, 1995.

[15]   J. Fresner, "Cleaner production as a means for effective environmental management", *J. Clean. Prod.,* vol. 6, no. 3-4, pp. 171-179, 1998.
[http://dx.doi.org/10.1016/S0959-6526(98)00002-X]

[16]   H. Freeman, T. Harten, J. Springer, P. Randall, M.A. Curran, and K. Stone, "Industrial pollution prevention! A critical review", *J. Air Waste Manage. Assoc.,* vol. 42, no. 5, pp. 618-656, 1992. [http://dx.doi.org/10.1080/10473289.1992.10467016]

[17]   Office of Air Quality Planning and Standards, *Development Of Air Pollution Control Cost Functions For The Integrated Iron and Steel Industry,* 1979.

[18]   US Environmental Protection Agency Staff and Risk Reduction Engineering Laboratory (US), *Pollution Prevention Research Branch, Facility pollution prevention guide.* Inspirations by Grace Lajoy, 1992.

[19]   University of Illinois, "Pollution Prevention: A Guide to Program Implementation", *Illinois Waste Management and Research Center, One E. Hazelwood Drive Champaign, Illinois 61820, TR-E09.,* 1993.

[20]   E. Phipps, *Pollution Prevention Concepts and Principles.* National Pollution Prevention Center for Higher Education, 1995.

[21]   Great Lakes National Program Office, *Promoting Pollution Prevention in the Great Lakes Basin,* 1998.

[22]   US EPA, *Guide to Industrial Assessments for Pollution Prevention and Energy Efficiency,* 2001.

[23]   R. Pojasek, and C. Metcalf, *An Organizational Guide to Pollution Prevention,* 2001.

[24]   M. Novotney, and R. Winer, *Municipal Pollution Prevention.* Good Housekeep. Pract. Cent. Watershed Prot. Ellicott Md, 2008.

[25]   S. Specktor, and N. Roy, "An Ounce of Pollution Prevention is Worth Over 167 Billion Pounds of Cure: A Decade of Pollution Prevention Results 1990-2000", *National Pollution Prevention Roundtable,* 2003.

[26]   "A Guide To Pollution Prevention Planning", *Texas Commissionon Environmental Quality Y.,* 2004.

[27]   Environmental Assistance Office, "Preventing Pollution During Vehicle Salvage", *Missouri Department of Natural Resources, PUB000394.,* 2004.

[28]   USEPA, *National management measures to control nonpoint source pollution from urban areas,* 2005.

[29]   Office of Pollution Prevention and Technical Assistance, "2005 Indiana Pollution Prevention Annual Report", *Indiana Department of Environmental Management, Indianapolis, IN, Government.,* 2006.

[30]   A. Kumar, N. Raman, and A. Vijayan, *2006 Guide to Pollution Prevention Tools,* 2006.

[31]   P. M. Shepard, "Advancing environmental justice through community-based participatory research", *Environ. Health Perspect.,* vol. 110, no. 2, pp. 139-139, 2002. [http://dx.doi.org/10.1289/ehp.02110s2139]

[32]   US EPA, "FY 2020 and FY 2021 Pollution Prevention Grant Program", *U.S. Environmental Protection Agency.,* 2020.

[33]   L.T.M. Bui, and S. Kapon, "The impact of voluntary programs on polluting behavior: Evidence from pollution prevention programs and toxic releases", *J. Environ. Econ. Manage.,* vol. 64, no. 1, pp. 31-44, 2012. [http://dx.doi.org/10.1016/j.jeem.2012.01.002]

[34]   United States Environmental Protection Agency, *Pollution Prevention and Toxics News Stories.*

[35]   US EPA, "Pollution Prevention Tools and Calculators", *Calculators to measure P2 environmental outcomes.* https://www.epa.gov/p2/pollution-prevention-tools-and-calculators (accessed May 06, 2020).

[36]   N. Hummel, and K. Davey, *Using EPA's P2 Cost & GHG Calculators to Measure Environmental Outcomes.*

[37]   A. Kumar, "Pollution Prevention Tools", *Air Pollution Research Group*. http://www.eng.utoledo.edu/ aprg/ppis/ppistools.htm (accessed Jan. 17, 2020).

[38]   L.N. Kuruppuarachchi, S.V.H. Madiraju, A. Kumar, and M. Franchetti, "Pollution Prevention Approaches to Analyze Assessment Data", *International Conference on Sustainable Solutions in Industrial Pollution, Water and Wastewater Treatment*, pp. 7-9, 2018.

[39]   M. Schratz, C. Gupta, T. Struhs, and K. Gray, "Reducing energy and maintenance costs while improving light quality and reliability with led lighting technology", *Conference Record of 2013 Annual IEEE Pulp and Paper Industry Technical Conference (PPIC)*, pp. 43-49, 2013. [http://dx.doi.org/10.1109/PPIC.2013.6656043]

[40]   F. Li, K. Wichmann, and R. Otterpohl, "Review of the technological approaches for grey water treatment and reuses", *Sci. Total Environ.*, vol. 407, no. 11, pp. 3439-3449, 2009. [http://dx.doi.org/10.1016/j.scitotenv.2009.02.004] [PMID: 19251305]

[41]   F.R. Spellman, *Handbook of water and wastewater treatment plant operations*. CRC press, 2013.

[42]   D. Pimentel, and M. Pimentel, "Global environmental resources *versus* world population growth", *Ecol. Econ.*, vol. 59, no. 2, pp. 195-198, 2006. [http://dx.doi.org/10.1016/j.ecolecon.2005.11.034]

[43]   J.G. March, M. Gual, and F. Orozco, "Experiences on greywater re-use for toilet flushing in a hotel (Mallorca Island, Spain)", *Desalination*, vol. 164, no. 3, pp. 241-247, 2004. [http://dx.doi.org/10.1016/S0011-9164(04)00192-4]

[44]   K.A. Mourad, J.C. Berndtsson, and R. Berndtsson, "Potential fresh water saving using greywater in toilet flushing in Syria", *J. Environ. Manage.*, vol. 92, no. 10, pp. 2447-2453, 2011. [http://dx.doi.org/10.1016/j.jenvman.2011.05.004] [PMID: 21621904]

[45]   J. Lambe, and R. Chougule, *Greywater-treatment and reuse*. IOSR J. Mech. Civ. Eng. IOSR-JMCE ISSN, 2008, pp. 2278-1684.

[46]   E. Friedler, R. Kovalio, and N.I. Galil, "On-site greywater treatment and reuse in multi-storey buildings", *Water Sci. Technol.*, vol. 51, no. 10, pp. 187-194, 2005. [http://dx.doi.org/10.2166/wst.2005.0366] [PMID: 16104421]

[47]   A. Campisano, and C. Modica, "Experimental investigation on water saving by the reuse of washbasin grey water for toilet flushing", *Urban Water J.*, vol. 7, no. 1, pp. 17-24, 2010. [http://dx.doi.org/10.1080/15730621003596739]

[48]   F. Razza, M. Fieschi, F.D. Innocenti, and C. Bastioli, "Compostable cutlery and waste management: An LCA approach", *Waste Manag.*, vol. 29, no. 4, pp. 1424-1433, 2009. [http://dx.doi.org/10.1016/j.wasman.2008.08.021] [PMID: 18952413]

[49]   S. Brown, "Greenhouse gas accounting for landfill diversion of food scraps and yard waste", *Compost Sci. Util.*, vol. 24, no. 1, pp. 11-19, 2016. [http://dx.doi.org/10.1080/1065657X.2015.1026005]

[50]   "Managing Food Waste for Sustainability", *Landfills versus Composting*. https://kb.wisc.edu/ dairynutrient/375fsc/page.php?id=48783 (accessed Jan. 18, 2021).

[51]   C. Bradford, M. Cain, M. Drumm, S. Hay, and P. Kay, *Biodegradable Alternatives for Styrofoam and Plastics*, 2003.

[52]   L. Liu, *Bioplastics in food packaging: Innovative technologies for biodegradable packaging* vol. 13. San Jose State Univ. Packag. Eng., 2006, pp. 1348-1368.

[53]   S. Velagapudi, A. Kumar, A. Spivak, and M. Franchetti, "Comparison of pollution prevention assessments for the facilities with and without energy star certification", *Environ. Prog. Sustain. Energy*, vol. 33, no. 4, p. n/a, 2014. [http://dx.doi.org/10.1002/ep.11935]

CHAPTER 9

Industrial Biofilter Design for Removal of Hydrogen Sulphide (H$_2$S) from Wastewater Treatment Plants

Zarook Shareefdeen[1,*]

[1] *Department of Chemical Engineering, American University of Sharjah, P.O. Box 26666, Sharjah, UAE*

Abstract: Hydrogen sulphide (H$_2$S) is the main odor-causing, toxic, and corrosive chemical found in wastewater treatment operations. Bio-oxidation based processes for air pollutant removal have become more attractive to the industry and numerous wastewater treatment facilities have replaced conventional air treatment technologies such as adsorption and chemical scrubbing with bio-oxidation based processes such as biofilters. Of the three main types of air phase bioreactors, biofilter is used more commonly than the others due to its simple configuration, ease of operation, and economic benefits. This chapter addresses challenges in the industrial biofilter design for H$_2$S removal from wastewater treatment facilities. Wastewater industry professionals, biofilter customers, biofilter vendors, and researchers who work in the field of odor and H$_2$S emission control and biofilter design will find this chapter very useful.

Keywords: Hydrogen sulphide (H$_2$S), Odor control, Biofilter design, Air pollutant, Wastewater treatment emissions.

INTRODUCTION

The wastewater at conventional treatment facilities comes from many sources with different wastes and flow characteristics. During the multi-step treatment process, the amount of pollutants in the water is reduced to permissible limits. The emission of hydrogen sulphide (H$_2$S) is prevalent at various concentration levels throughout the wastewater treatment processes including preliminary treatment, primary sedimentation, biological treatment, secondary sedimentation, and sludge processing. Carrera-Chapela *et al.* (2014) report that odor contribution due to H$_2$S emissions at wastewater treatment facilities is in the following order: sludge stor-

---

[*] **Corresponding author Zarook Shareefdeen:** Department of Chemical Engineering, American University of Sharjah, P.O. Box 26666, Sharjah, UAE; Tel: 971-6-515-2988; E-mail: zshareefdeen@aus.edu

G. Venkatesan, S. Lakshmana Prabu and M. Rengasamy (Eds.)

age (26%), sludge thickening (19%), sludge dewatering (17%), aerated grit chamber (13%), primary sedimentation (11%), aeration tank (5%), secondary sedimentation (5%), and bar screen (4%).

H_2S is a toxic and corrosive pollutant associated with several health effects including headaches, nausea, eye irritation, paralysis, and even death when exposed to high concentration levels. Wastewater treatment plant operators are recurrently exposed to H_2S. In a recent study, H_2S exposure levels are mapped to sewer works with regard to the types of jobs, tasks, seasons, and geographical locations (Austigard *et al.*, 2018). The odor threshold of H_2S is in the range of 0.7–200 $\mu g/m^3$ and it is recommended to maintain the levels at or below this limit to avoid complaints from neighbouring communities (Godoi *et al.*, 2018). H_2S emission is also a big nuisance due to its rotten egg odor characteristics; thus, strict regulations are in effect for the removal of this highly odorous and toxic gas at wastewater treatment facilities.

Conventional treatment methods such as condensation, carbon adsorption, chemical oxidation, scrubbing and incineration can be used for controlling air emissions (Barbosa and Stuetz, 2013); however, there are many disadvantages of these technologies which include high disposal cost, chemical use, and energy cost. The condensation is economical for only high boiling point compounds and more concentrated (*i.e.* high level of pollutants) air streams. The adsorption process is effective when the concentration of the pollutant is low. Spent adsorbents after several regenerations become solid waste that needs to be landfilled or incinerated. In scrubbing, pollutants are absorbed into scrubbing chemicals.

Chemical costs are high and the generated liquid waste requires further treatment. Incineration or thermal oxidation is widely used by many industries due to its high efficiency. However, due to high-energy usage (or fuel consumption), incineration is not economical when concentrations levels are low. Furthermore, the highest amount of greenhouse gases (GHGs) such as carbon dioxide and NO_X is produced by incineration. Biological processes also can produce greenhouse gases such as N_2O and CH_4. Furthermore, chemical oxidation and incineration are less effective when H_2S is present at low ppm (parts per million) levels in the contaminated airstreams.

In the last two decades, bio-oxidation based process units such as biofilters, bio-scrubbers and bio-trickling filters have become more attractive to industry (Rabbani *et al.*, 2016). Hence, these units have replaced conventional technologies for air treatment at numerous wastewater treatment facilities. In a review paper, Mudliar *et al.* (2010) report that the biological waste air treatment using bioreactors has increased popularity in the control of volatile organic compounds (VOCs) and odors, since they offer a cost-effective and environment-friendly alternative to conventional air pollution control methods.

Of the three types of bio-reactors mentioned, biofilter is used more widely than the others due to its simple configuration, ease of operation and economic benefits (Barbosa and Stuetz, 2013; Shareefdeen and Singh, 2005). Other biological methods such as activated sludge diffusion (ASD) used for H_2S removal has limited application (Barbosa and Stuetz, 2013). The objective of this chapter is to provide practical guidelines based on sound knowledge of the industrial biofilter design for the removal of H_2S from wastewater treatment facilities.

Industrial Biofilter Design

Biofilter Configurations

Biofiltration takes place in an air-phase biological unit in which an air stream contaminated with pollutant is passed through a bed of media particles which are placed in a housing in the shape of a rectangular or circular tank or in a custom-made concrete structures depending on the size of airflow volume treated. As the air passes through a bed of medium, the pollutant such as H_2S, is transferred from the air phase to biofilms which are formed on the interior and exterior surfaces of the packing media particles. The metabolism of the pollutants by bacteria requires a moist environment; thus the moisture is provided by saturating the incoming air and by an occasional spray of water on top of the biofilter media. One of the important pre-treatment units for biofiltration is the humidifier. A simple configuration of a biofilter system is shown in Fig. (**1**). In biofilter design, many factors need to be considered.

Airstream Characterizations

Airstream needs to be characterized to find out the types of contaminants, concentration levels, temperature, humidity, and particulate levels (*i.e.*, dust, aerosols) and other contaminants in the air streams. Based on the air stream characterization, one needs to determine if pre-treatment steps are required. If non-biodegradable compounds are present and they are odorous or toxic, these compounds need to be removed by physical or chemical methods through chemical scrubbers or other applicable air pollution control devices. It is important to consider that the selection of pre-treatment methods does not interfere overall biofiltration process.

Biofiltration is effective and economical for biodegradable odor-causing compounds such as hydrogen sulphide (H_2S) and VOC contaminants at low concentration levels (*i.e.* ppm range). Highly soluble and low molecular weight VOCs (*i.e.* methanol, ethanol, aldehydes, acetates, ketones and some aromatic hydrocarbons) and inorganic compounds (*i.e.* hydrogen sulphide, ammonia) are easily biodegradable in biofilters but low molecular weight aliphatic hydrocarbons such as methane, and some chlorinated compounds are more difficult to be removed.

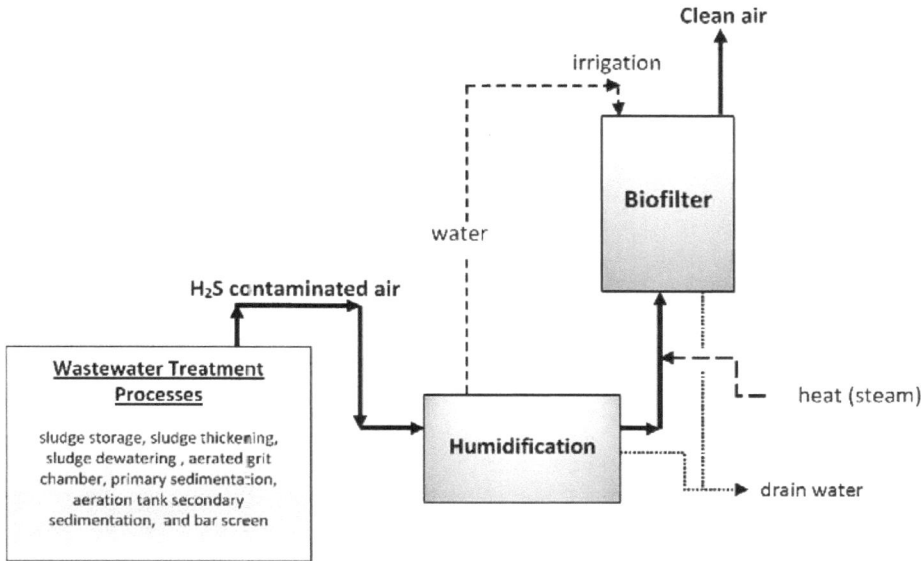

Fig. (1). Schematic diagram of a biofilter describing biofiltration process used for removal of H_2S.

Humidification of Inlet Air

Inlet air streams must be humidified so that the media is moist and the bacteria growth can be established within the biofilter; if the media is not moist, part of the bed will be unused and the residence time will be below the design value leading to poor removal. Occasional water spray on top of the media is also essential to keep the media moist, in addition to the pre-humidification step. Humidification of the inlet air by passing the air through a water tank or by heating the water in a tank may be acceptable for lab-scale systems but such methods of humidification are not suitable for full-scale applications especially when large airflow volumes are treated. Bacterial growth, scale deposits, and plugging of spray nozzles are some of the common problems encountered in humidification units.

Modulating Flow Rates

In order to determine the dimension of a biofilter unit, the total incoming airflow rate based on a number of processes that contribute to odor and air exchanges must be determined. The inlet airflow rate must be adjusted so that the concentration levels are acceptable for the biofiltration process. If the concentration level is high, dilution of the incoming air may be required. Shareefdeen (2012) presented a simple software tool that allows for estimating dilution volume and water demand for humidification units based on the psychrometric calculations.

It is economical to keep the empty bed residence time (EBRT) low, which is defined as the media volume to the volumetric airflow rate. Although high EBRT of 3-4 minutes can provide over 99% removal of a pollutant in lab-scale experiments, it is not practical and economical for industrial applications because of its large footprint or biofilter size requirements. Research that focuses on the high removal of a pollutant but disregards the practical range of EBRT values cannot be scaled up. More than 30 seconds of EBRT for H_2S removal using a biofilter is not generally recommended for economic reasons. EBRT can be determined based on the literature data, laboratory or pilot-scale experiments and through predictive modeling. The EBRT is very much dependent on the type of media used, microorganisms present in the media and the kinetics of bio-degradation. Microorganisms used in biofilters are aerobic and typically, biofilters are inoculated with activated sludge obtained from a wastewater treatment plant or microbes taken from a biofilter that has been used for the treatment of similar compounds. However, other organisms such as fungi are also used in biofilters for the removal of H_2S (Hansen *et al.*, 2018). Once the EBRT is known, the physical dimensions of the biofilter unit can be determined easily based on the factors such as the airflow volume, available footprint for construction, climate condition and geographical location.

Series-Parallel and Modular Biofilter Units

Biofilter units are built as a single, dual, up flow, down flow, series, parallel, modular or custom-built units. A single bed is suitable for smaller airflow volume and the down flow mode facilitates washing of the bed to remove acidic by-products formed due to bio-oxidations of H_2S. A series of biofilter units can be used by a single unit if removal efficiency is not sufficient and parallel beds allow for treating large airflow volumes. Many biofilter vendors sell modular biofilter units; however, when treating large air flow volumes (*i.e.*, 500,000 cubic feet per minute, cfm), biofilters are usually custom-built.

Treating H₂S when Mixed with other Pollutants

Biofilter designs may vary due to different compounds present in the airstreams and specifications received from biofilter customers. For example, a typical customer A (please refer to Table **1**) requires a biofilter for H_2S removal for a flow rate that varies between 3500 cfm - 4900 cfm. In addition to H_2S, other Volatile Organic Compounds (VOCs) including acetone, butanone, and toluene are also present in the airstreams; thus pre-treatment or post-treatment are required for removal of these VOCs to meet environmental regulations for HAP (hazardous air pollutants) in addition to H_2S removal. Furthermore, H_2S concentration is not steady but fluctuates and peaks up to 30 ppm.

Table 1. Customer # A- specifications.

Flow	3500-4900 cfm (cfm = cubic feet/min.)
Compounds	mainly H_2S, others acetone, butanone, toluene
Concentration (avg.)	15 ppm
Concentration (peak)	30 ppm
Operation	flow rate varies by 1400 cfm

Similarly, customer # B requires (Table **2**) a biofilter for treating 16,000 cfm of air containing H_2S from an air stripper. Stripping efficiency and liquid phase concentration are available but the concentration of H_2S in the design specification is missing or not available. Thus, design requirements from customers vary and additional basic engineering applications and knowledge on air pollution control devices for VOC control are required for providing a complete solution and accurate biofilter design for H_2S removal.

Table 2. Customer # B- specifications.

Flow	16, 000 cfm
Compounds	H_2S
Concentration in the water	4 mg/liter
Stripper efficiency	95%+ (forced air aerator)
Water flow rate to the stripper	8.0 million gal/day

Air stream characterization and knowledge of the biodegradability of the compounds present are important in the biofilter design. In order to provide accurate designs and guarantees for complete removal of H_2S for biofilter units, sales engineers and professionals should acquire knowledge on the type of compounds or specific conditions present in the air streams; otherwise, it will be costly to biofilter customers as well as vendors who supply biofilters.

Biofilter Evaluation Parameters

In evaluating biofilter performance, percent removal efficiency (RE %), elimination capacity (EC) or removal rate (RR) and mas mass loading (load) are used. In the evaluation of RE%, removal due to physical processes such as adsorption by the media, absorption by the moisture, *etc.* must be excluded.

During the start-up of a biofilter, higher RE% is observed and this is due to physical processes; thus one may be misled to conclude high efficiency is due to the efficiency of the media or microbes tested. RE% during the process start-up is not only due to the biofiltration process but also due to absorption into moisture as well as adsorption onto solid media particles and biofilms. Similarly, during the process shut down, RE% may reduce due to desorption effects. A more accurate value for RE% can be obtained only after a complete steady state is reached.

When a third party evaluates a biofilter for removal performance guaranty, one must make sure, the operation is under steady state and there is no dilution air coming into the biofilter systems through leaks in the ducts or biofilter vessels. Elimination capacity (EC), also known as removal rate (RR), is defined as the mass of contaminants removed per volume of biofilter media per time. Mass loading (load) is defined as the mass of pollutants entering a biofilter per volume of media per time. Mass loading becomes equal to EC when the contaminant is completely (RE=100%) degraded. EC *versus* load curve is useful in determining the efficiency of biofilter media.

Biofilter Media Selection

Several types of media materials are used in the laboratory, pilot and commercial biofilters. The most common packing type is the organic media such as wood bark, peat, soil, compost, *etc*. Although, significant success is achieved with organic media, degradation of the media occurs within a few years of operation; thus organic media is not promising for a long term operation. For every 3-4 years, organic media need to be replaced. Hou *et al.* (2016) studied the biofiltration of H_2S with two types of compost; bio-dehydration stage-compost and curing stage-compost. According to their study, curing stage compost had a higher capability for H_2S removal. The economic constraints compel customers to select low-cost organic biofilter media such as compost. Several biofilter vendors through research and development have produced synthetic media which are resistant to degradation over time (Herner and Shareefdeen, 2008). For the removal of H_2S from biogas, Su *et al.*, (2013) used light-expanded clay aggregates and hollow spherical polypropylene balls as the bioreactor media. A recent study

(Erdirencelebi and Kucukhemek, 2018) reports that H_2S can be oxidized by using Fe^{3+} and it can be a component in the biofilter media preparation.

The selection of biofilter packing media is critical for effective performance. The media selection is based on several factors including the ability to support bacterial growth, large surface area, cost of raw materials for the media preparation, pressure drop consideration, void fraction and media life. Packing media must provide good absorption capacity for moisture retention, adsorption property to handle concentration fluctuations during shock loadings, pH buffering capacity to mitigate acid and sulphur accumulation within the media, good pore structure and surface area for biofilm growth, and structural property to avoid compaction over time.

Accumulation of sulphur in the media bed will create biofilter operational upsets. From a previous study, Fig. (**2**) illustrates SEM images of fresh and used biofilter media which are covered with sulphur deposits after a long period of use (Shareefdeen *et al*., 2003a). Such sulphur deposits reduce the surface area available for biomass growth; thus reduction in the removal of H_2S will occur. Acidic byproducts must be washed out of the media.

Fig. (2). Used biofilter media (right) is compared with the fresh media (left). Sulphur deposits are shown in the used biofilter media. (SEM image, adapted Fig. 10 of Shareefdeen *et al*., 2003a with permission from John Wiley, Copyright 2021).

When structural properties of the media are not considered in the selection process, the washing of sulphur deposits from the media or neutralizing the media for pH adjustments will create excessive pressure drops and channeling problems. This will cause a significant drop in performance efficiency. Often media with only good biodegradable properties are selected without considering important physical characteristics and compression strength of the media. Due to this reason, poor performance may be expected when scaled up from lab-scale to industrial-scale units.

BIOFILTER MONITORING

After commissioning of the biofilter systems, regular monitoring of the media is an important part of biofilter operational practice. Plant growth, fungus development, clumping of the media, sulphur accumulation, worm growth within organic media, *etc*. are also expected if biofilter media is not monitored and maintained well. Typical media analysis should include general observation, particle size-distribution analysis, moisture, pH, microbial count and nutrient analysis. Shareefdeen *et al*. (2002) report bark media analysis of a biofilter that had been operated for 4 years. Media analysis was performed at different locations. Moisture content, pH and total microbial content varied from 59.1% 64.1%, 4.93-6.96 and 1.7×10^6 to 4.4×10^6 CFU/g, respectively. Shareefdeen *et al*. (2003a) analyzed in the media and media leachate as 2310 ppm and 1460 ppm, respectively. Cudmore and Gostomski (2005) report particle size distribution on a weight basis for bark media as follows: 8 mm (22%), 5.6 mm (48.9%), and 3.3 (18.7%). Thus, changes in particle size distribution over time and visual observation provide information on media deterioration and bed compaction issues. Nutrient analysis is done to make sure macro- and micro- nutrients are available for bacterial growth. The total microbial count provides information on acceptable levels of bacteria. Moisture content and changes in pH levels affect the removal efficiency; thus, these parameters are monitored periodically.

Heat Integration During Winter Operation

Biological oxidation is an exothermic process, however the amount of heat released due to biofiltration is not substantial since inlet concentration to the biofilter is low or in the ppm range. However, as the temperature increases, the biological oxidation rate increases. Therefore, maintaining the temperature of operation at optimum growth conditions (*i.e.* 25°C) is important. When biofilter is operated in very cold climate conditions, design considerations should include circumventing any process upsets due to temperature change. Since incoming air is humidified, during winter operations, the water in the inlet airstream condenses in the biofilter and forms ice; thus, the incoming air should be heated to ensure no freezing of the media or blockage of any drain lines at the bottom of the biofilter vessels (Shareefdeen and Hashim, 2009). If the drain line freezes, then the media will be flooded with water; this will cause the shutting down of the biofilter operation altogether. In some cases, insulation of the ducts may be required to avoid freezing. The addition of steam to inlet air stream is an option to maintain an acceptable temperature within the biofilter media. In a published work

(Shareefdeen, 2012), a simple design tool based on pyschrometric calculation is incorporated into predictive software to estimate heat requirements. Based on this tool, steam generator ratings can be calculated for winter operations of industrial biofilters. As part of the process units, steam generators are used at many biofilter locations where winter temperatures can go down to -30°C or lower. When incoming air is heated with the steam, the humidification step may be avoided.

Pressure Drop and Operational Cost

The project cost of a biofilter is typically dependent on the blower cost and power requirements to operate the blower. Thus, maintaining a low-pressure drop across the blower fan, air ducts and the media is important. Blower can be installed either upstream or downstream of the biofilter. Since H_2S is an odorous gas, it is recommended to operate the biofilter under a vacuum and place the blower at the downstream of the biofilter. This will ensure that no leaks occur from the ducts or biofilter vessels to the surroundings so that the odor releases can be minimized.

Increase in the media pressure drop could result in channeling through the media and requires more power for the blower; thus increasing the electrical costs associated with the operation of the blowers. The pressure drop of a biofilter bed may be estimated using the well-known Ergun Equation (Dorado *et al*., 2010).

$$\frac{\Delta P}{L} = 150 \frac{\mu U}{d_p^2} \frac{(1-\varepsilon)^2}{\varepsilon^3} + 1.73 \frac{(\rho U)^2}{d_p} \frac{(1-\varepsilon)}{\varepsilon^3} \qquad (1)$$

In addition to the parameters included in the above Equation **1**, changes in the moisture content, biomass accumulation, type and shape of the particles also affect pressure drop across the media. Therefore, Equation **1** gives only an estimate. Laboratory or pilot tests may be required to establish correlations that can more accurately predict pressure drops across the biofilter. Letto *et al*. (2007) compared the pressure drop of a biofilter with organic media over a 3-year period and they noticed a significant increase in the pressure drop. Over the 3 years, pressure drop through the media bed increased from about 150 to 450 Pa/m. Thus, accurate estimation of pressure drop using Equation **1** is impractical since void fraction, particle size or shape of the particles change over a period of time. However, this may not be the case when biofilter media having good structural properties are used.

BIOFILTER MODELS AND PERFORMANCE

Steady-state biofiltration models along with lab-scale or pilot performance data on H_2S removal are often used to determine EBRT which helps in designing biofilter systems. Ottengraf and van den Oever's (1983) model is one of the most commonly used mathematical models in biofilter design. This model assumes plug flow in the air phase through a biofilter bed, negligible gas-biofilm mass transfer resistances and existence of equilibrium at the interface of the air-biofilm. Pollutants in the airstreams diffuse into the biofilm through air/biofilm interface and get biodegraded. Biodegradation kinetics without any oxygen-limitations is assumed and the kinetics follow either zero-order or first-order.

There are plenty of research works done to extend Ottengraf and van den Oever (1983) model and the details of these models are described in the literature (*i.e.*, Deshusses and Shareefdeen, 2005). Because of analytical presentation of the models and simplicity, Ottengraf and van den Oever (1983) model continues to play an important role in the biofilter design.

Ottengraf and van den Oever's (1983) first order kinetic model is more suitable when concentration levels are low and zero order kinetic is applicable at higher values of H_2S. At high H_2S concentration levels, overall process can be limited by diffusional mass transfer. Thus, for zero-order kinetics, two models (reaction limited, diffusion limited) exist. H_2S performance data that gives the best fit to one of the models is used in determining EBRT and sizing biofilters. Depending on the microbes and the type of media used, zero or first order kinetic model may be used for sizing biofilter units after validating the model with the lab or pilot scale experimental data.

Zero-order reaction-limited model:

$$\frac{C_{Out}}{C_{in}} = 1 - \frac{K_0 A_S \delta\, EBRT}{C_{in}} \tag{2}$$

where, K_o: zero-order rate constant ($kg.m^{-3}.s^{-1}$); A_S: biofilm surface area ($m^2.m^{-3}$); δ: biofilm thickness (m); C_{in} and C_{out}: concentrations at the inlet and outlet of the biofilter ($kg.m^{-3}$);

Zero-order diffusion-limited model:

$$\frac{C_{out}}{C_{in}} = \left(1 - EBRT\, A_S \sqrt{\frac{K_0 D_e}{2mC_{in}}}\right)^2 \tag{3}$$

where m: air/biofilm distribution coefficient (-) and D_e: effective diffusivity in the biofilm ($m^2.s^{-1}$).

First Order Model:

$$\frac{C_{out}}{C_{in}} = \exp\left(\frac{-EBRTA_sD_e}{m\delta}\phi\tanh\phi\right) \tag{4}$$

where, K: first-order rate constant and parameter (s^{-1}), is defined as (-) $\sqrt{\frac{K}{D_e}}$.

The literature shows the applicability of these models in designing industrial scale biofilters for H$_2$S removal. In a study (Shareefdeen *et al.*, 2003a), lab scale columns were packed with synthetic media particles. Compositions of the media particles are described in the US Patent No. US 2008/0085547 (Herner and Shareefdeen, 2008). H$_2$S removal performance data that were obtained from the lab scale and several field scale biofilter systems were compared with the three mathematical models described above. For this study, performance data followed closely zero-order diffusion limited model.

$$\sqrt{\frac{C_{H2S,air}}{C_{H2S,air-inlet}}} = \left\{1 - \alpha_{lump}EBRT\sqrt{\frac{1}{C_{H2S,air-inlet}}}\right\} \tag{5}$$

In the above equation, the parameter ^{a}lump was found to be 0.18 when H$_2$S is in ppm$_v$ unit and EBRT is in seconds. Based on this model, sizing of modular biofilter units was found and used in the industrial scale. Typical size of modular biofilter units based on these models is given in Table **3**.

Table 3. Modular biofilter specifications.

	EBRT		Typical Dimensions			
-	30sec.	20 sec.	Diam. (in)	W(ft)	H (ft)	L (ft)
			Flow Rate (CFM)			
1	330	550	23	-	6	-
2	1056	1584	-	8	8	12
3	1760	2640	-	8	8	20
4	2816	4224	-	8	8	32
5	3520	5280	-	8	8	40

In another study (Shareefdeen *et al.*, 2011), using a light weight building material, hollow cylindrical particles were made, coated with nutrients, and subsequently used as biofilter media. The media was then packed in a laboratory scale biofilter and placed on a mobile cart. This unit was then placed at a wastewater pumping station for field data collection. H_2S performance data were collected at different empty bed residence time (EBRT). When compared with all the models, for this case, the first order model (Equation **4**) predicted H_2S removal performance data well according to first order relation. Fig. (**3**) shows the model predicted and experimental data at 30 seconds EBRT (Shareefdeen *et al.*, 2011). This model equation can be used in biofilter sizing for various inlet concentrations levels of H_2S provided the same media type is used as packing materials in the biofilter.

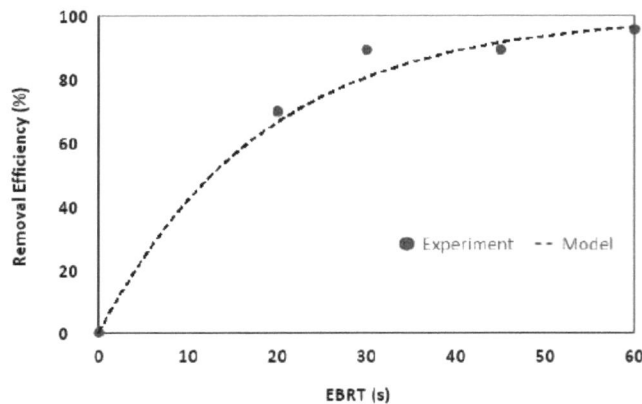

Fig. (3). Comparison between the model and the experimental data (Shareefdeen *et al.*, 2011).

Effect of Other Reduced Sulphur Compounds

Although, biofilters can be successfully designed and employed for removal of H_2S, often other reduced sulphur compounds such as methyl mercaptan, dimethyl sulfide and odor causing VOCs are also emitted in the wastewater treatment process. Thus, even after complete removal of H_2S, odor issues and odor complaints from nearby community may persist.

Furthermore, in some regions, H_2S levels can be extremely high; in these cases, alternate biological treatment systems such as a bio-scrubber, bio-trickling filter *etc.* along with physical treatment systems may be required.

CONCLUSION

In this work, air stream characterization, selection of biofilter configurations, humidification methods, heat demand calculations for winter operation, biofilter media selection, parameters that affect media performance, sulphur deposits problems, theoretical model equations, empty bed residence time evaluation using model fitting and case studies that demonstrate the concepts are discussed. Furthermore, limitations of the biofilter process is highlighted. Although there are several advantages of using biofilters for controlling air emissions, one should know clearly the limitations of this technology before implementation. Some limitations include: (a) large footprint requirement, (b) only biodegradable pollutants can be treated, (c) concentrations should be relatively low, and (d) organic media needs to be replaced more frequently. Biofilter vendors may experience disappointments when extending biofilter success stories of North American markets to the other regions of the globe where extreme hot weather, high levels of H_2S and varying humidity conditions prevail. Thus, further development, technological and design changes in biofilter systems are required to meet such challenges. Wastewater industry professionals, biofilter customers, biofilter vendors and researchers who work in the field of odor and H_2S emission control and biofilter design will find this chapter very useful.

CONSENT FOR PUBLICATION

Not applicable.

CONFLICT OF INTEREST

The author declares no conflict of interest, financial or otherwise.

ACKNOWLEDGEMENT

Declared none.

BIBLIOGRAPHY

[1]   A. Austigard, K. Svendsen, and K. Heldal, "Hydrogen sulphide exposure in waste water treatment", *J. Occup. Med. Toxicol.,* vol. 13, no. 1, pp. 1-12, 2018.

[2]   V.L. Barbosa, and R.M. Stuetz, "Performance of activated sludge diffusion for biological treatment of hydrogen sulphide gas emissions", *Water Sci. Technol.,* vol. 68, no. 9, pp. 1932-1939, 2013.

[3]   F. Carrera-Chapela, A. Donoso-Bravo, and J.A. Souto, "Modeling the odor generation in WWTP: An integrated approach review", *Water Air Soil Pollut.,* vol. 225, pp. 1932-1942, 2014.

[4]   R. Cudmore, and P. Gostomski, Biofilter design and operation for odor control.*Biotechnology for Odor and Air Pollution Control.* Springer Publishing: Germany, 2005.

[5]   M.A. Deshusses, and Z. Shareefdeen, Modeling of biofilters and bio-trickling filters for odor and VOC control applications.*Biotechnology for Odor and Air Pollution Control.* Springer Publishing: Germany, 2005.

[6]   A.D. Dorado, J. Lafuente, D. Gabriel, and X. Gamisans, "The role of water in the performance of biofilters: Parameterization of pressure drop and sorption capacities for common packing materials", *J. Hazard. Mater.,* vol. 180, no. 1–3, pp. 693-702, 2010.

[7]   D. Erdirencelebi, and M. Kucukhemek, "Control of hydrogen sulphide in full-scale anaerobic digesters using iron (III) chloride: performance, origin and effects", *Water S.A.,* vol. 44, pp. 1-20, 2018.

[8]   A. Godoi, A. Grasel, G. Polezer, A. Brown, S. Potgieter-Vermaak, D. Scremim, I. Yamamoto, and R. Godoi, "Human exposure to hydrogen sulphide concentrations near wastewater treatment plants", *Sci. Total Environ.,* vol. 610-611, pp. 583-590, 2018.

[9]   M.J. Hansen, C.L. Pedersen, L.H.S. Jensen, L.B. Guldberg, A. Feilberg, and L.P. Nielsen, "Removal of hydrogen sulphide from pig house using biofilter with fungi", *Biosyst. Eng.,* vol. 167, pp. 32-39, 2018.

[10] B.H. Herner, and Z. Shareefdeen, "Biofilter media and systems and methods of using same to remove odor causing compound from waste gas streams", *Patent No. US 2008/0085547 A1,* vol. April 10, 2008.

[11] J. Hou, M. Li, and T. Xia, "Simultaneous removal of ammonia and hydrogen sulfide gases using biofilter media from the bio-dehydration stage and curing stage of composting", *Environ Sci Pollut,* vol. 23, pp. 20628-20636, 2016.

[12] D. Letto, D. Webb, and M. Rupke, "Herner B. A comparison of an organic biofilter and inorganic biofilter for the treatment of residual odors emanating from a bio-solids dewatering facility", *Proceedings of the Water Environment Federation,* pp. 709-725, 2007.

[13] S. Mudliar, B. Giri, K. Padoley, D. Satpute, R. Dixit, P. Bhatt, R. Pandey, A. Juwarkar, and A. Vaidya, "Bioreactors for treatment of VOCs and odours – A review", *J. Environ. Manage.,* vol. 91, 2010.10391054

[14] S.P.P. Ottengraf, and A.H.C. van den Oever, "Kinetics of organic compound removal from waste gases with a biological filter", *Biotechnol. Bioeng.,* vol. 25, pp. 3089-3102, 1983.

[15] K.A. Rabbani, A. Charles Kayaalp, R. Cord-Ruwisch, and G. Ho, "Pilot-scale biofilter for the simultaneous removal of hydrogen sulphide and ammonia at a wastewater treatment plant", *Biochem.,* vol. 107, p. 110, 2016.

[16] Z. Shareefdeen, B. Herner, and S. Wilson, "Biofiltration of nuisance sulfur gaseous odors from a meat rendering plant", *Chemical Tech. and Biotech.,* vol. 77, no. 12, pp. 1296-1299, 2002.

[17] Z. Shareefdeen, B. Herner, and D. Webb, "Wilsong, S. Hydrogen sulfide (H_2S) removal in synthetic media biofilters", *Environ. Prog.,* vol. 22, no. 3, pp. 1-12, 2003.

[18] Z. Shareefdeen, B. Herner, D. Webb, and S.S. Wilsong, "Synthetic media biofilter eliminates hydrogen sulphide and other reduced sulphur compounds odors", 76th WEFTEC-2003 Conference; Los Angeles CA, 2003b, October 11-15.

[19] Z. Shareefdeen, and A. Singh, *Biotechnology for Odor and Air Pollution Control.* Springer Publishing: Germany, 2005.

[20] Z. Shareefdeen, and M. Hashim, "Winter operation of biofilters for hydrogen sulphide removal", *Int. J. Chem. React. Eng.,* vol. 7, no. 1, pp. 1-12, 2009.

[21] Z. Shareefdeen, W. Ahmed, and A. Aidan, "Kinetics and modeling of H_2S removal in a novel biofilter", *Advan. in Chem Engg and Science.,* vol. 1, pp. 72-76, 2011.

[22] Z. Shareefdeen, "A biofilter design tool for hydrogen sulfide removal calculations", *Clean Technol. Environ. Policy,* vol. 14, p. 543, 2012.

[23] J.J. Su, Y. Chang, Y.J. Chen, K.C. Chang, and S.Y. Lee, "Hydrogen sulfide removal from livestock biogas by a farm-scale bio-filter desulfurization system", *Water Sci. Technol.,* vol. 67, no. 6, pp. 1288-1293, 2013.

[24] M. Zilli, C. Guarino, D. Daffonchio, S. Borin, and A. Converti, "Laboratory-scale experiments with a powdered compost biofilter treating benzene-polluted air", *Process Biochem.,* vol. 40, no. 6, pp. 2035-2043, 2005.

CHAPTER 10

Assessment of Ground Water Quality Using GIS Techniques

B.P. Naveen[1,*] and **K.S. Divya**[2]

[1] *Department of Civil Engineering, Amity University, Haryana, India*

[2] *VTU Extension Center, Karnataka State Remote Sensing Applications Centre, Bangalore, India*

Abstract: In the present study, the groundwater quality was tested around the Gandhinagar sub-watershed covering a neighborhood of 53.63 sq. km, which lies between north latitudes 12°46′ and 13°58′ and east longitudes 77°21′ and 78°35′ within the state of Karnataka, India. For the study, data collection includes maps, toposheets, water quality data, well locations, village locations, *etc.* The above-said data has been collected from various government departments of Karnataka. After the data collection, the base map was prepared using ArcMap. The water samples have been tested and then used as an attribute database to design thematic maps showing various water quality parameters.

Keywords: Arc GIS, Groundwater quality, Quality parameter, Thematic maps.

INTRODUCTION

In the current scenario, in most of the cities in India, the water demand is met by groundwater utilization, as the surface water is either polluted or deficient. Groundwater is the primary source in India for domestic use and agriculture and the industrial sector (Umamaheswari *et al.*, 2015) [1]. In the present scenario, 85% of household water requirements in rural areas, 55% of farmers' irrigation water requirements, 50% of domestic water requirements in urban areas, and 50% of process water requirements of industries are met by groundwater. Groundwater has been tapped for the past twenty years because of the increasing demand for water and water resources management. This leads to water scarcity. The groundwater level has been falling rapidly day by day. It is essential to start investigations oriented toward groundwater quantification and qualification, which can become the basis to form plans for its exploitation, management, and conservation.

---

* **Corresponding author B.P. Naveen:** Department of Civil Engineering, Amity University, Haryana, India; E-mail: bp.naveen864@gmail.com

G. Venkatesan, S. Lakshmana Prabu and M. Rengasamy (Eds.)

Freshwater sources include lakes, rivers, and groundwater. Groundwater serves as the primary water source in the urban environment, used for drinking, industrial and domestic purposes. In the past few decades, increasing anthropogenic activities, especially industrialization, have become a threat to humankind and the ecosystem. Improper waste management practices often lead to the degradation of groundwater, with attendant health and environmental implications. It was initially believed that the soil and sediment layers deposited above an aquifer acted as a natural filter that kept pollutants from the surface from infiltrating down to the groundwater. However, it has become widely understood that those soil layers often do not adequately protect aquifers, and groundwater is inherently susceptible to contamination from anthropogenic activities. Remediation is costly and sometimes not practical.

Generally, physiochemical and biological analyses assess groundwater quality (Fatombi *et al.*, 2012; Kulandaivel *et al.*, 2009; Senthilkumar and Meenambal, 2007) [2 - 4]. Hydrochemical analysis of groundwater can also be utilized for groundwater quality (Ranjan *et al.*, 2013) [5]. The geographic Information System (GIS) mapping technique is used for groundwater assessment and its utilization for drinking, irrigation, and construction needs (Ravikumar *et al.*, 2013) [6]. An estimate can be obtained to better understand groundwater by representing the data by ArcGIS Software (Thiyagarajan & Baskaran, 2013) [7]. Over some time, there is a possibility of groundwater quality changes due to hydrology and geologic conditions (Pandey and Tiwari, 2009) [8].

A study on groundwater quality analysis was carried out for Coonoor taluk in the Nilgiris district. A survey was conducted on Ground Water Quality mapping in the Municipal Corporation of Hyderabad using GIS techniques. There is a need for a specific strategy and guidelines that would concentrate on a particular part of groundwater management, which means protecting groundwater from contamination. This study aims to visualize the spatial variation of specific physicochemical parameters through GIS. This work's main objective is to assess groundwater quality using GIS based on the available physicochemical data from 14 locations in the Gandhinagar sub-watershed area. Quality maps were created to visualize, analyze, and understand the relationship between the measured points.

MATERIALS AND STUDY AREA

Study Area: Kolar, the golden city of the Indian state of Karnataka, is the headquarters of the Kolar district. It is located in the State's southern region and is the easternmost district of Karnataka State. The neighborhood is bounded by the Bangalore Rural district in the west, Chikballapur district in the north, Chittoor District of Andhra Pradesh in the east, and south Krishnagiri and Vellore district of Tamil Nadu. The Kolar district receives an average annual rainfall of 748 mm, and the mean daily minimum and maximum temperatures are 22.7°C - 38°C, respectively. The sub-watershed area covers around 25 villages. Gandhinagar sub-watershed is undulating too plain. The northern and eastern parts of the world, forming the valley of Palar Basin, are well cultivated. The overall elevation varies from 849 to 1130 m above the mean water level. The study area has different landforms, like hills, ridges, pediments, plains, and valleys (Fig. **1**).

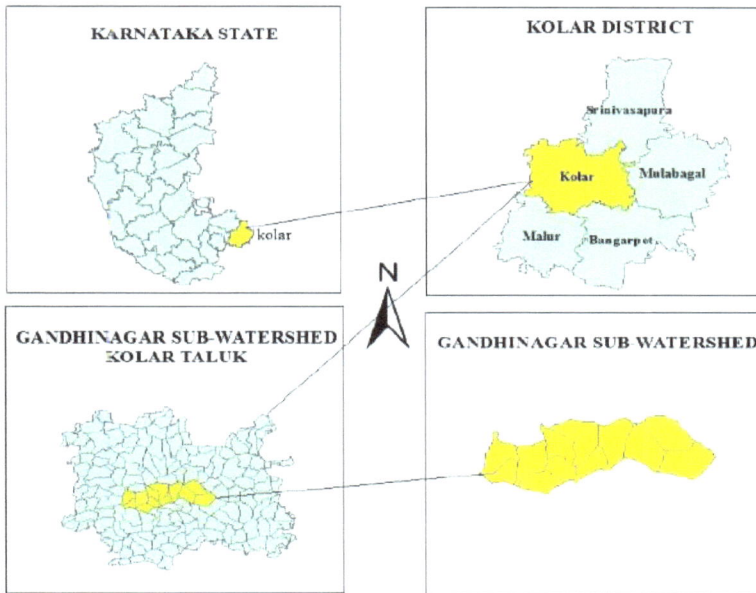

Fig. (1). Location map of gandhinagar sub-watershed.

SAMPLE COLLECTION AND ANALYSIS

The study area's base map was prepared using Survey of India topographic sheets (57G/15, 57G/16) and digitized using ArcGIS 9.3 software. Fourteen groundwater samples (Bore wells) were collected in April 2015. The groundwater samples were collected from 14 different bore wells in the Gandhinagar sub-watershed (Fig. **2**) in pre-cleaned, sterilized polyethylene bottles. The utmost care was taken to fill the bottles without air bubbles at each sampling site. The collected samples were labelled and transported to the laboratory using a refrigerator box. The reagents used in experimentation were prepared by using double distilled water.

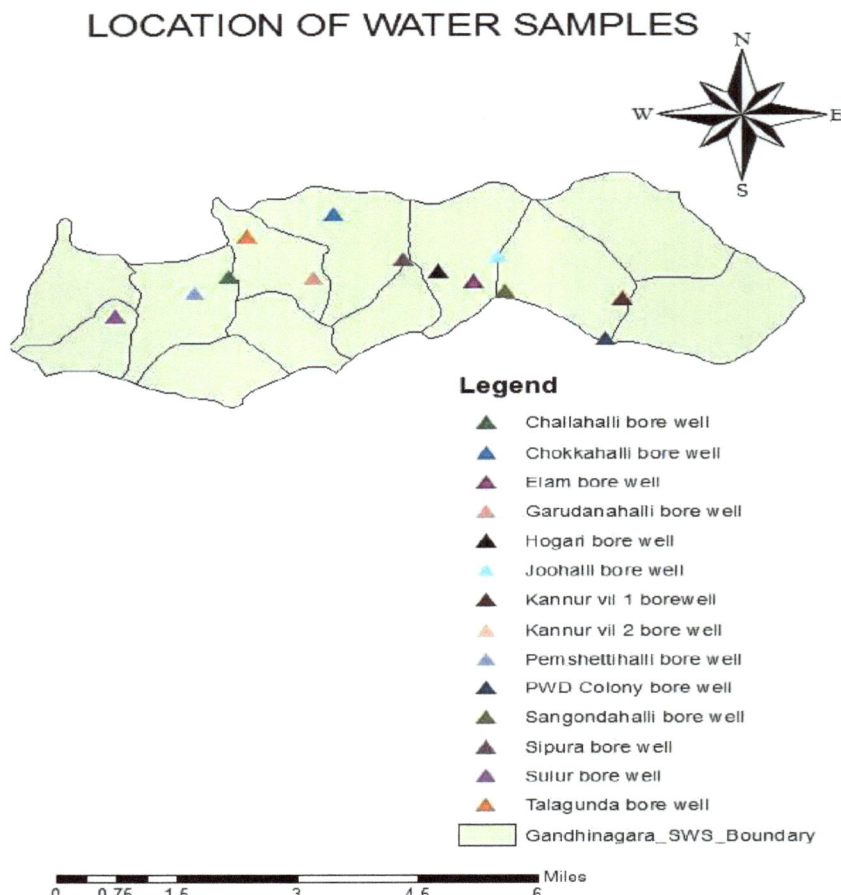

Fig. (2). Water sample locations using GPS.

Physicochemical parameters and ionic parameters were carried out according to standard methods for examining water and wastewater unless otherwise stated (APHA, 1998) [9]. The analytical techniques adopted for the analysis of water quality parameters are presented in Table **1**.

Table 1. Methods of analysis for different parameters of leachate.

Parameters	Instrument Used to Identify the Parameters
pH	pH meter
TDS, mg/l	TDS meter

(Table 1) cont.....

Chloride, mg/l	Argentometric Titration
Calcium, mg/l	EDTA titration
Alkalinity, mg/l	EDTA titration
Total alkalinity, mg/l	Titrimetric
Total hardness,mg/l	Titrimetric
Magensium, mg/l	Titrimetric
Fluoride, mg/l	Spectrophotometer
Turbidity, NTU	Nephelometric Turbidity
Heavy metal	Atomic Absorption Spectrophotometer (AAS)

Satellite data: Satellite remote sensing provides multi-spectral, multi-spatial, multi-temporal data useful for resource inventory, monitoring, and management. Satellite data is amenable to both visual interpretation and digital analysis. Within the present study, visual performance was carried out to organize different thematic maps of the study area. The image obtained from resourcesat for the year 2012 is employed within the study, as shown in Fig. (**3**).

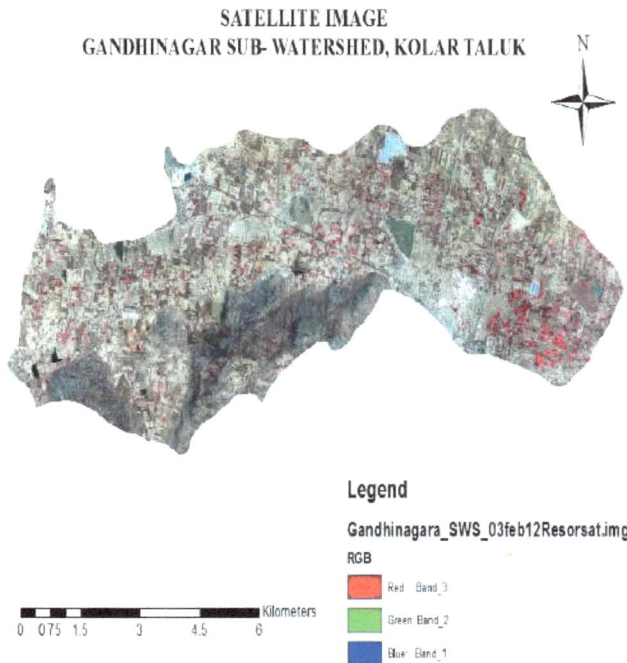

Fig. (3). Satellite image of Gandhinagar sub watershed.

METHODOLOGY

The below flow chart shows the method followed to study water quality in the Gandhinagar subwatershed area (Fig. **4**).

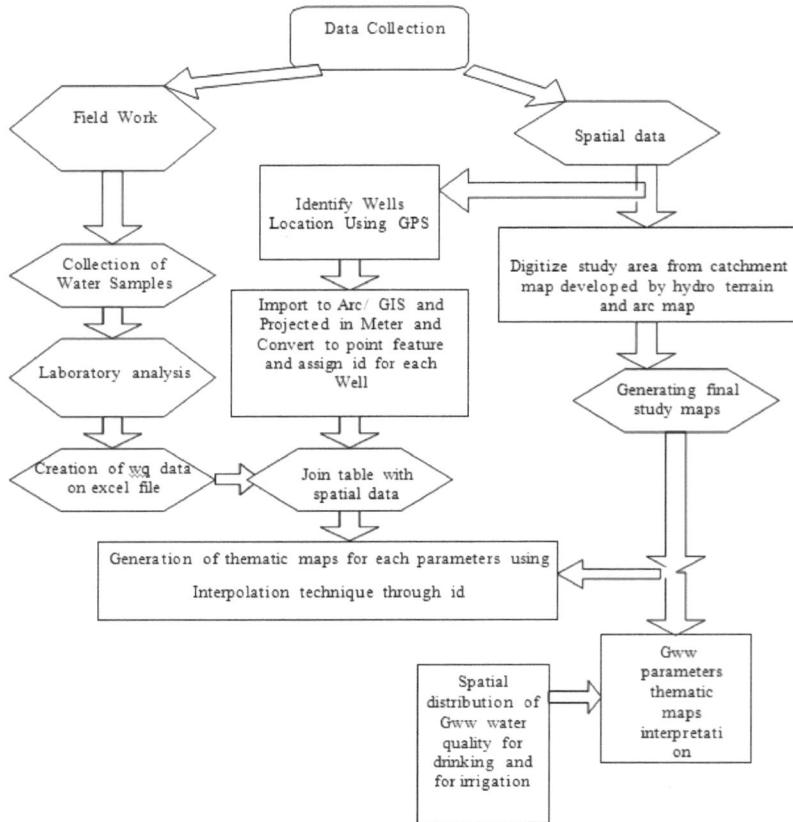

Fig. (**4**). Flowchart of the methodology.

RESULTS AND DISCUSSIONS

The collected fourteen groundwater samples were analyzed for various physicochemical parameters and compared with BIS and WHO standards Table **2** and the exact values of all parameters are presented in Table **3**. Thematic maps for the parameters of pH, Total Alkalinity (TA), Total Hardness (TH), chloride(Cl⁻), fluoride (F⁻), Magnesium (Mg^{2+}), Calcium (Ca^{2+}), Turbidity, and Iron (Fe) were integrated. The spatial integration for groundwater quality mapping was carried out using ArcGIS spatial analyst extension. It can be seen in the final drinking water quality maps that a large area on the east, west, and northeast parts of the study area has water potable in the absence of better alternate sources.

Table 2. Groundwater data in comparison with BIS and WHO.

Parameters	BIS (10500:2012)		WHO (2011)	Samples
	Acceptable Limit	Permissible Limit		
pH	6.5-8.5	-	7.5-8.5	6.23-7.99
Turbidity, NTU	5-10	5	-	0-0.01
TH (mg L^{-1})	200	600	200	152-800
TA (mg L^{-1})	200-600	-	120	260-660
Cl$^-$ (mg L^{-1})	250	1000	250	150-332
Ca^{2+} (mg L^{-1})	75	200	75	200-840
Mg^{2+}(mg L^{-1})	30	No relaxation	50	0-48
F$^-$ (mg L^{-1})	1	1.5	0.5-1.5	2-8
Fe (mg L^{-1})	0.3	No relaxation	< 0.3	0.1-0.3

To assess groundwater usability for purposes such as drinking, irrigation, *etc.*, the results obtained were compared with Indian standards IS 10500 (BIS, 1991) [10] and World Health Organization standards (WHO, 2011) [11]. Water consumption for drinking purposes must be free from physical parameters such as color, odor, *etc.* Also, to maintain proper irrigation practices, water quality, soil types, and cropping practices play an essential role as the excessive quantity of dissolved ions in water affects habitats such as plants, soil, and productivity. These effects lower the osmotic pressure in plant cells and decrease metabolic activity (Ravikumar and Somashekar, 2013) [6].

Table 3. Physicochemical characteristics of Gandhinagar sub watershed area.

Location	pH	TA	TH	Ca^{2+}	Mg^{2+}	F$^-$	Turbidity	Cl$^-$	Fe
PWD Colony	7.58	660	800	840	0	2	0	275	0.1
Kodi Kannur	7.95	480	720	696	24	4	0.01	272	0.2
Kannur	7.9	460	700	690	10	4	0.01	150	0.3
Sangondahalli	7.1	400	368	320	48	6.4	0	290	0.2
Joohlli	7.44	360	200	416	0	6.4	0.01	283	0.1
Elam	6.49	320	184	464	0	6.4	0	310	0.1
Hogari	6.87	400	344	384	0	8	0	230	0.1
Sipura	6.9	380	224	200	24	6.4	0	294	0.2
Garudanahalli	6.38	340	376	800	0	8	0	330	0.1

(Table 3) cont.....

Chokkahalli	7.47	340	152	264	0	4	0	332	0.1
Talagunda	6.62	380	288	461	0	4	0	303	0.3
Challahalli	6.23	500	328	336	0	8	0.01	270	0.1
Pemshettihalli	7.99	260	192	328	0	8	0	225	0.2
Sulur	7.51	340	216	480	0	4	0	260	0.3

SPATIAL DISTRIBUTION

The spatial distribution of the Physicochemical analysis of groundwater samples collected from 14 different bore wells in the Gandhinagar sub-watershed is shown in Figs. (**5-13**). pH value denotes the acidic or alkaline condition of water which is expressed on a scale ranging from 0 to 14, which is the common logarithm. The recommended pH range for treated drinking water is 6.5 to 8.5. It can be observed that from Fig. (**5**) that the optimum pH required within the range from 6.23-7.99. The maximum permissible limit for drinking water, as given by BIS, is 8.5. If pH is less than acceptable, water takes acidic nature, and it may cause tuberculation in water supply systems. This shows that the groundwater of the study area is not acidic.

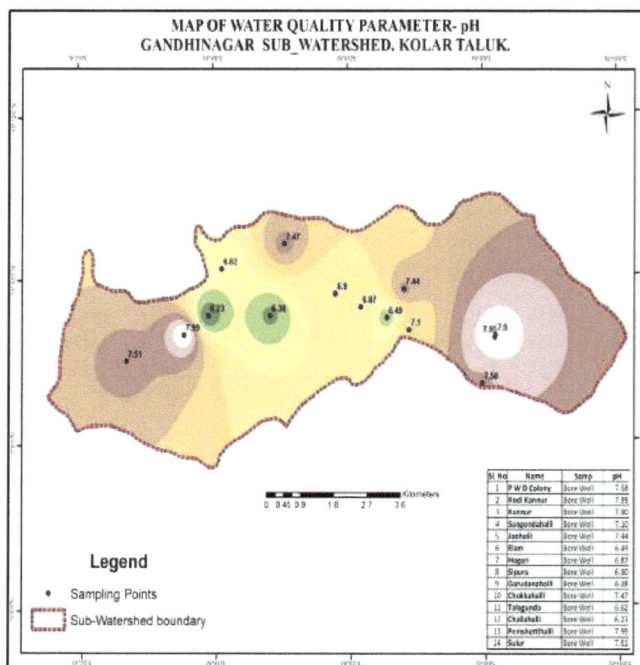

Fig. (5). Map of water quality parameter with respect to pH.

The alkalinity of water is a measure of its capacity to neutralize acids. It is expressed as mg/l in terms of calcium carbonate. Alkalinity is an important parameter in evaluating the optimum coagulant dosage. From the parametric analysis, total alkalinity was in the range of 200-600mg/l (Fig. **6**). All the samples got permissible alkalinity levels.

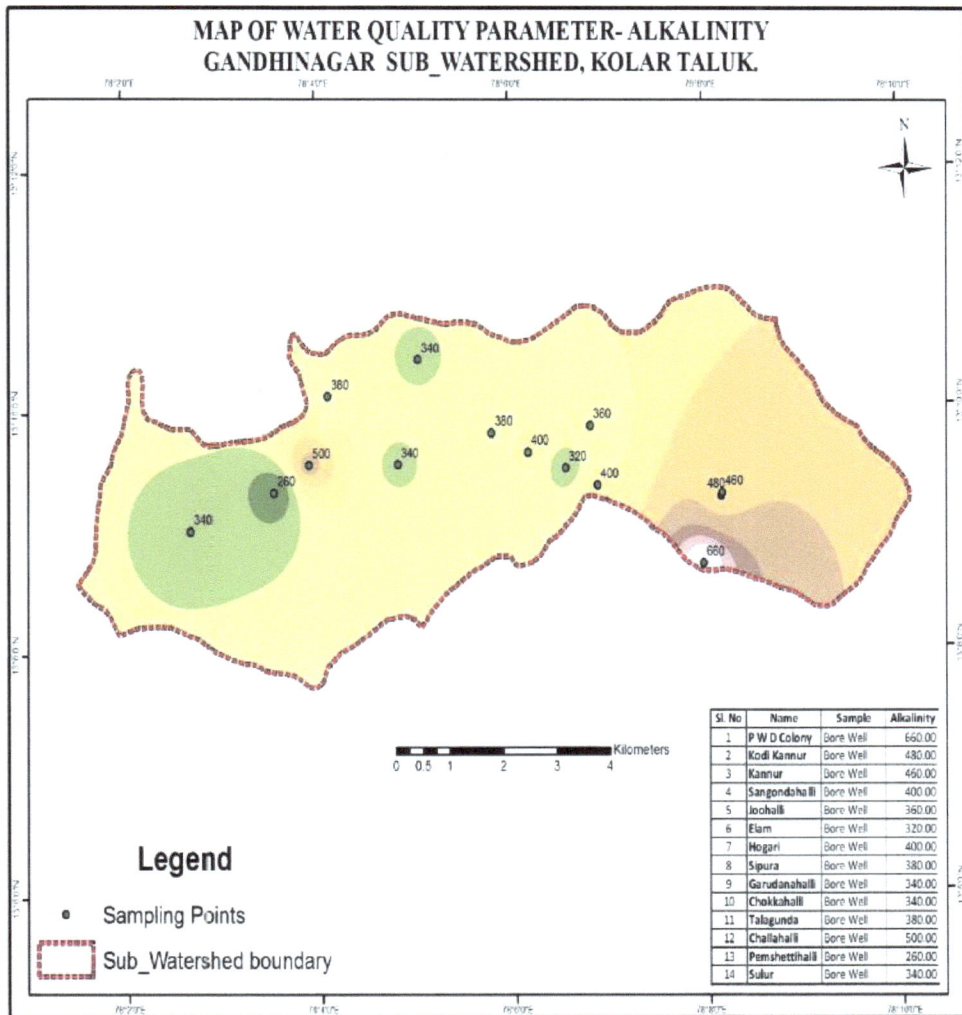

Fig. (6). Map of water quality parameter with respect to total alkalinity.

Total Hardness of water signifies them dissolved mineral content. If water consumes excessive soap to produce lather, it is said to be hard. Hardness is cause by divalent metallic actions. The principal hardness causing cations are calcium, magnesium, strontium, ferrous and manganese ions. The total hardness of water is defined as the sum of calcium and magnesium concentrations, both expressed as calcium carbonate, in mg/L. Temporary or carbonate hardness can be precipitated by prolonged boiling. Non-carbonate ions cannot be precipitated or removed by boiling, hence the term permanent hardness. Total hardness was in the range of 152-800 mg/l (Fig. 7). for 3 locations that exceeded the allowable limits as per the BIS and WHO standards. In general, a high level of hardness level makes water not potable and causes the scaling problem (Naveen *et al.*, 2017) [12].

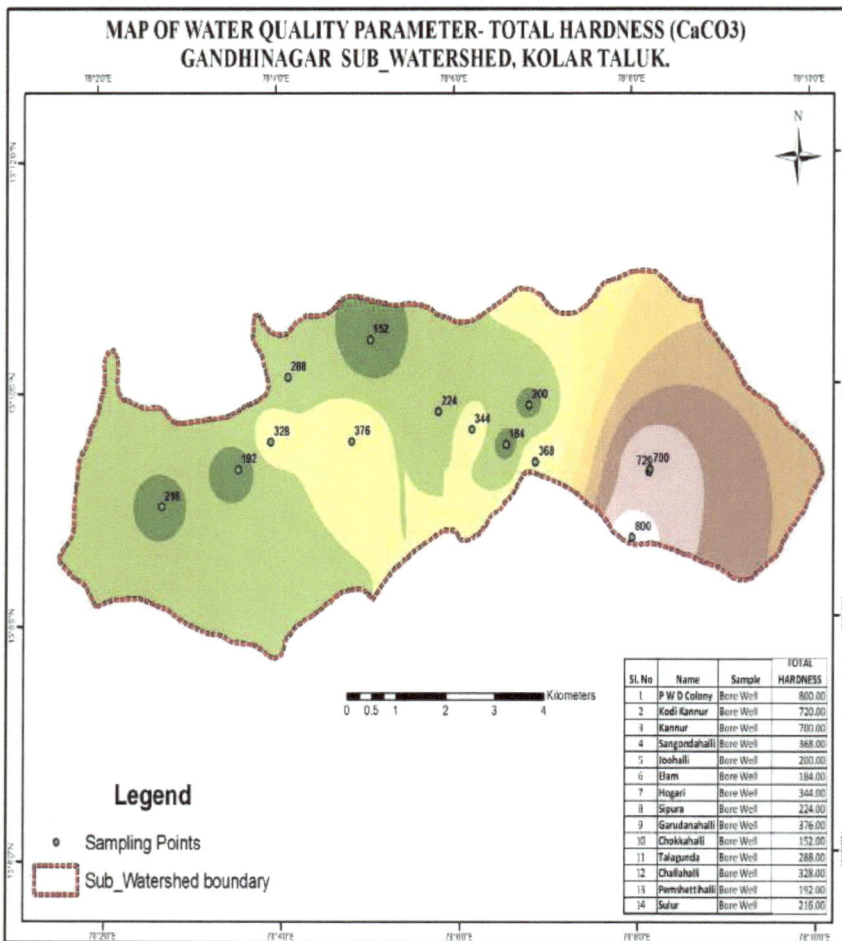

Fig. (7). Map of water quality parameter of total hardness.

Turbidity is the cloudiness or haziness caused by individual particles which makes the water appear non-transparent. If a large amount of suspended solids are present in water, it will appear turbid in appearance indicating presence of impurities. Turbidity was in the range of 0.01 NTU. The permissible turbidity levels were in the range of 5-10 NTU (Fig. **8**).

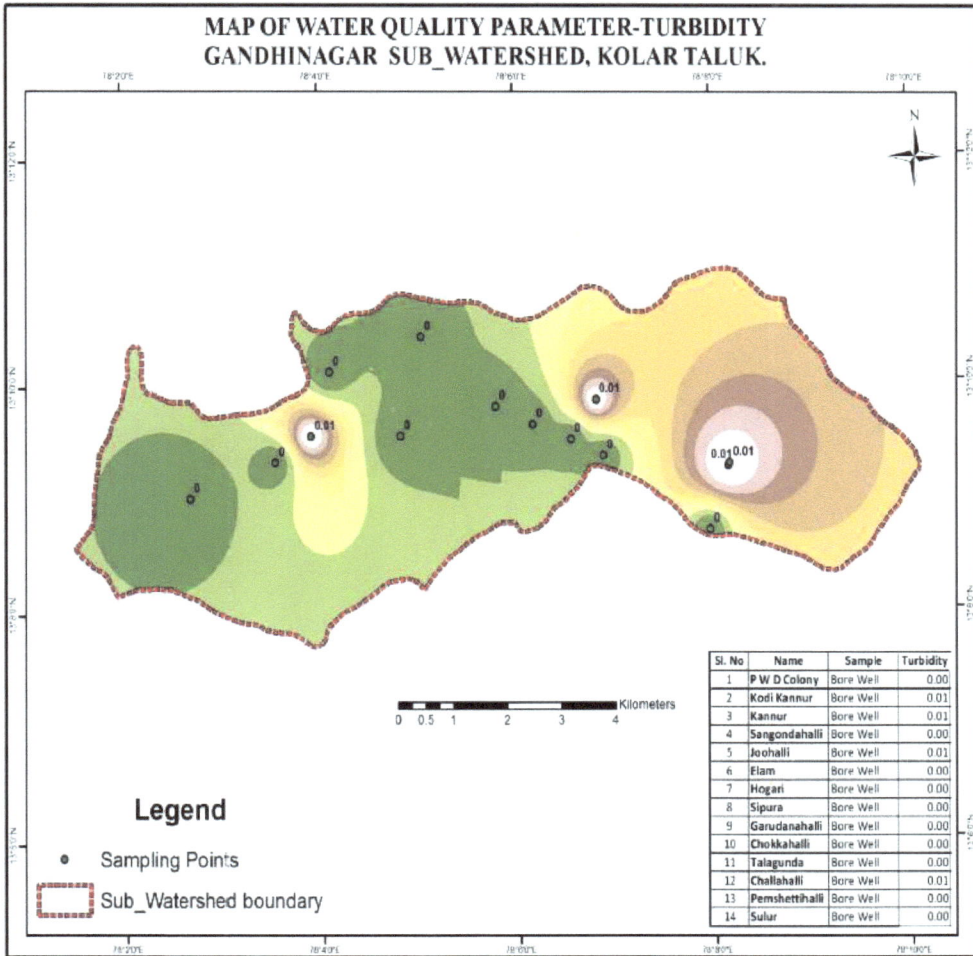

MAP OF WATER QUALITY PARAMETER-TURBIDITY GANDHINAGAR SUB_WATERSHED, KOLAR TALUK.

Sl. No	Name	Sample	Turbidity
1	P W D Colony	Bore Well	0.00
2	Kodi Kannur	Bore Well	0.01
3	Kannur	Bore Well	0.01
4	Sangondahalli	Bore Well	0.00
5	Joohalli	Bore Well	0.01
6	Elam	Bore Well	0.00
7	Hogari	Bore Well	0.00
8	Sipura	Bore Well	0.00
9	Garudanahalli	Bore Well	0.00
10	Chokkahalli	Bore Well	0.00
11	Talagunda	Bore Well	0.00
12	Challahalli	Bore Well	0.01
13	Pemshettihalli	Bore Well	0.00
14	Sulur	Bore Well	0.00

Legend

•   Sampling Points

▭   Sub_Watershed boundary

Fig. (8). Map of water quality parameter with respect to turbidity.

Anionic concentrations such as calcium (Fig. **9**) were in the range of 200-840 mg/l, which revealed that almost all the sampling points exceeded the desirable limit as the desirable limit of calcium is 75mg/l. The permissible limit is 200 mg/l as per BIS standards. The formation of calcium in water is mainly due to minerals like limestone, dolomite, gypsum, *etc.* Insufficiency of calciumcauses severe rickets; excess causes concretionsin the body such as kidney or bladder stones andirritation in urinary passages (CPCB 2008) [13].

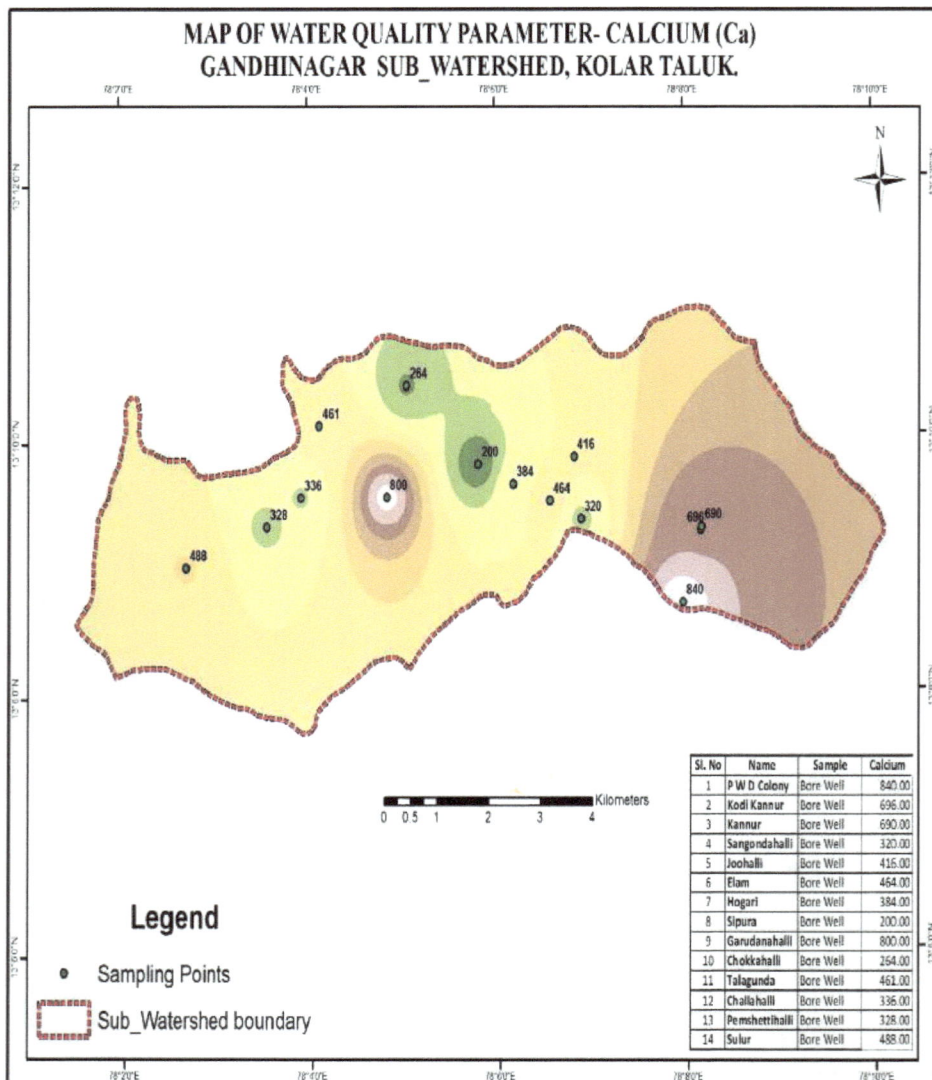

Fig. (9). Map of water quality parameter with respect to calcium.

Similarly,magnesium is present in the groundwaterfrom natural sources like granitic terrain whichcontain large concentration of these elements. The geochemistry of the rocktypes may have an influence in the concentration ofMg in groundwater. High concentration of magnesiummay cause laxative effect (CPCB, 2008) [13]. Magnesium (Fig. **10**) was in the range of 48 mg/l, which reduces the soil quality, reduces the crop yield, and gives toxicity when it exceeds 50% of the magnesium ratio (Ramkumar *et al.*, 2013) [14]. According to the BIS standard, magnesium's desirable limit is 30mg/l and no relaxation for its permissible limit.

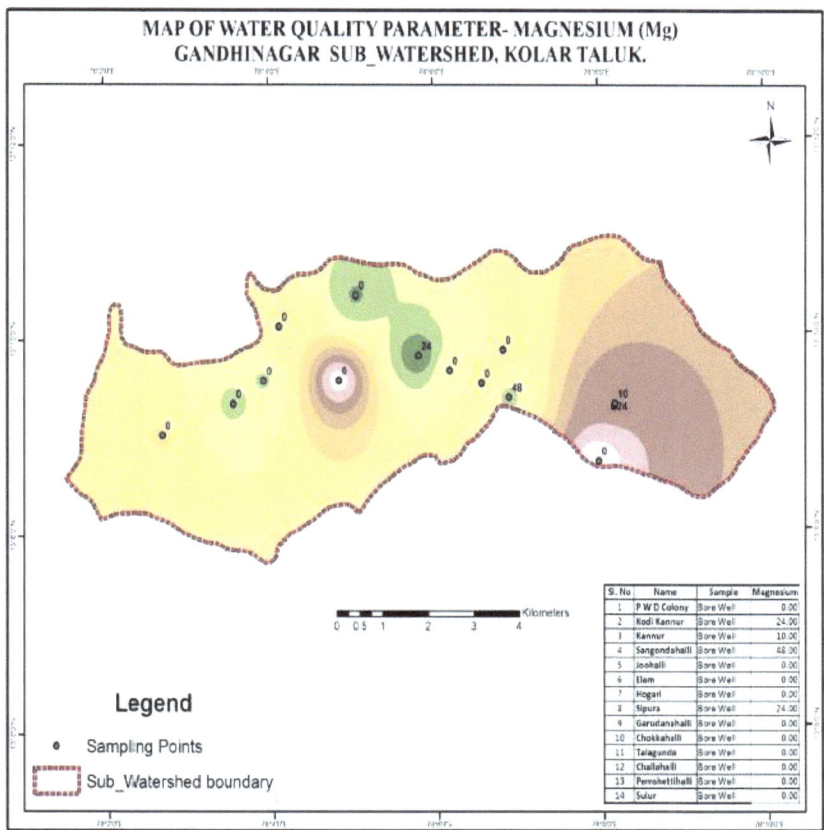

Fig. (10). Map of water quality parameter with respect to magnesium.

Similarly, chloride ion may be present in combination with one or more of the cations of calcium, magnesium, iron and sodium. Excessive chloride in water indicates presence of septic tank effluents, animal feeds, industrial effluents, irrigation drainage, and seawater intrusion in coastal areas. The desirable limit of chlorides is 250 mg/l, and the permissible limit is 1000 mg/l as per BIS standards. The chlorides for the study area ranged from 150-332 mg/l (Fig. **11**). It can be noted that the chloride content of 11 places (out of 14) was exceeded the prescribed limit.In natural waters, the probable sources ofchloride comprise the leaching of chloride-containingminerals (like apatite) and rocks with which the watercomes in contact, inland salinity and the dischargeof agricultural, industrial and domestic wastewaters (Abbasi, 1998) [15].

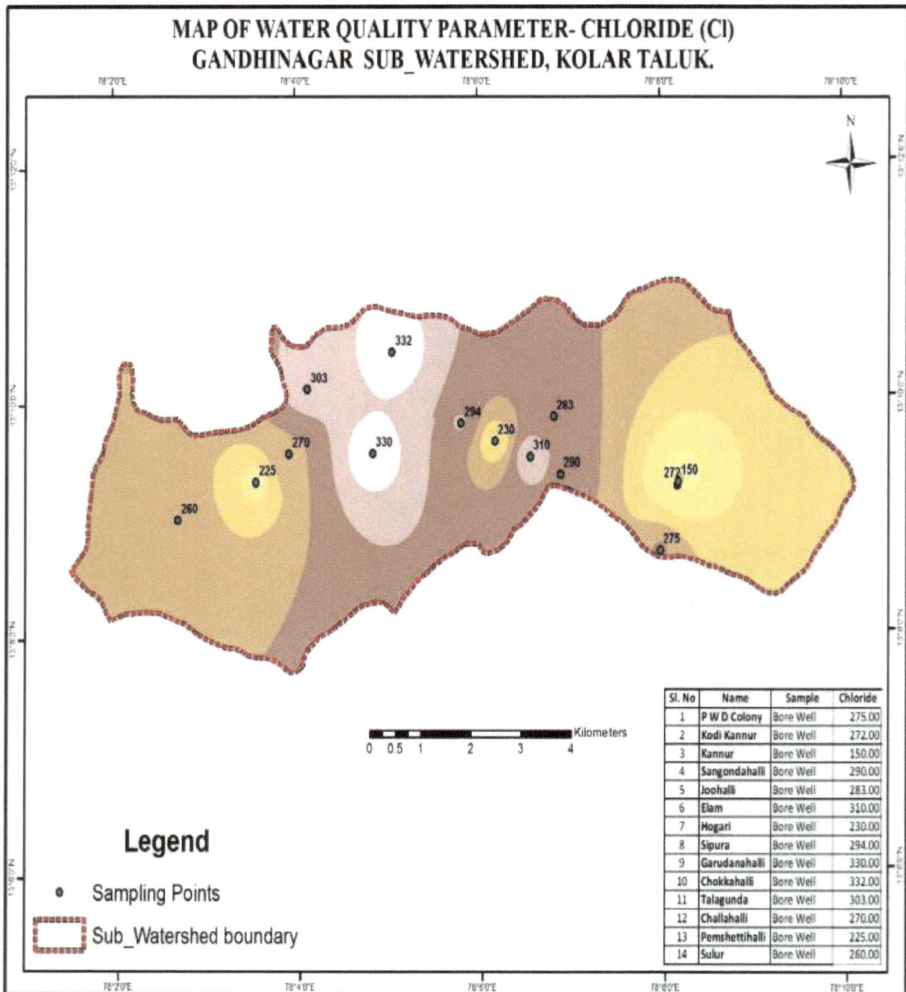

Fig. (11). Map of water quality parameter with respect to chloride.

Likewise, fluoride is a naturally occurring compound derived from fluorine. It is found in many rocks and minerals in the soil and enters drinking water as water passes through these soils. Fluoride has been shown to prevent tooth decay, but too much fluoride can cause teeth discoloration. Fluoride concentrations were in the range of 2-8mg/l (Fig. **12**). The desirable limit of fluoride is 1.5.

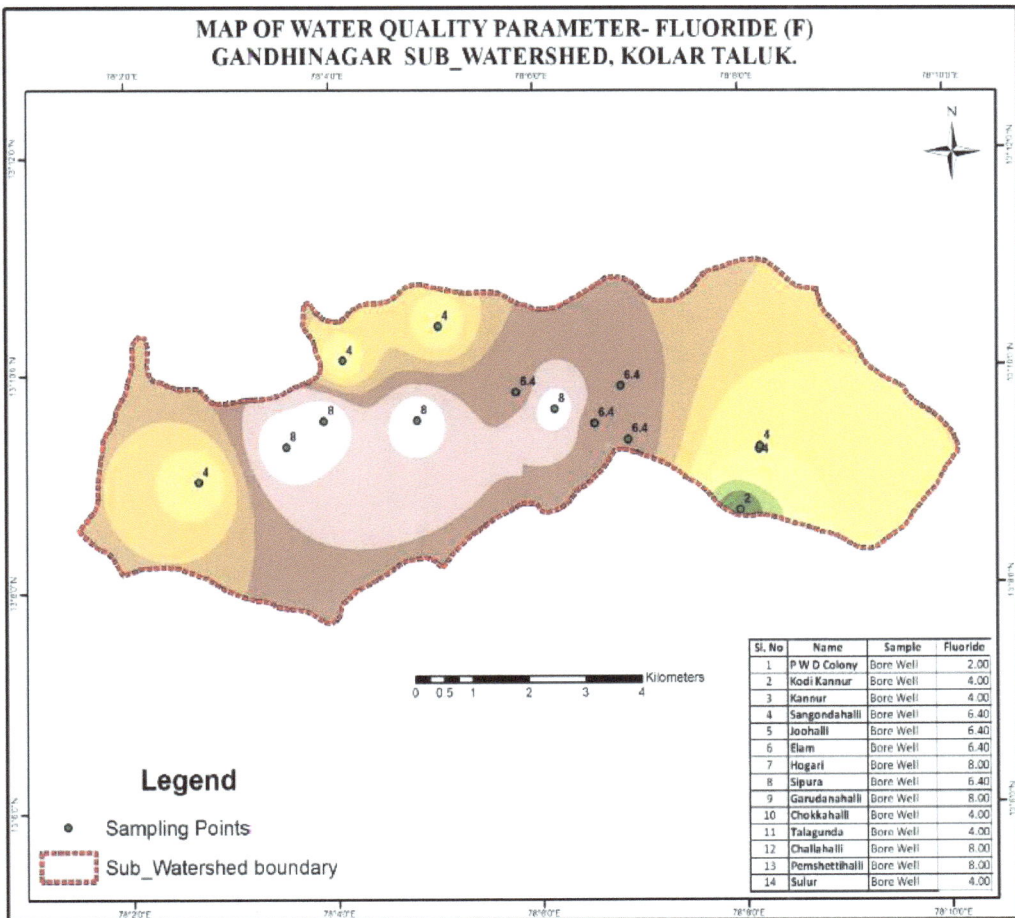

MAP OF WATER QUALITY PARAMETER- FLUORIDE (F) GANDHINAGAR SUB_WATERSHED, KOLAR TALUK.

Sl. No	Name	Sample	Fluoride
1	P W D Colony	Bore Well	2.00
2	Kodi Kannur	Bore Well	4.00
3	Kannur	Bore Well	4.00
4	Sangondahalli	Bore Well	6.40
5	Joohalli	Bore Well	6.40
6	Elam	Bore Well	6.40
7	Hogari	Bore Well	8.00
8	Sipura	Bore Well	6.40
9	Garudanahalli	Bore Well	8.00
10	Chokkahalli	Bore Well	4.00
11	Talagunda	Bore Well	4.00
12	Challahalli	Bore Well	8.00
13	Pemshettihalli	Bore Well	8.00
14	Sulur	Bore Well	4.00

Legend

• Sampling Points

Sub_Watershed boundary

Fig. (12). Map of water quality parameter with respect to fluoride.

Similarly, Iron is found on earth mainly as insoluble ferric oxide. When it comes in contact with water, it dissolves to form ferrous bicarbonate under favorable conditions. This ferrous bicarbonate is oxidized into ferric hydroxide, which is a precipitate. Iron imparts bad taste to the water and incrustations in water mains. Iron concentrations were in the ranges of 0.1-0.3mg/l (Fig. **13**). The desirable limit of Iron is 0.3mg/l, and no relaxation for its permissible limit. The Iron occurs naturally in the aquifer, but groundwater levels can be increased by dissolving the ferrous borehole and hand pump components.

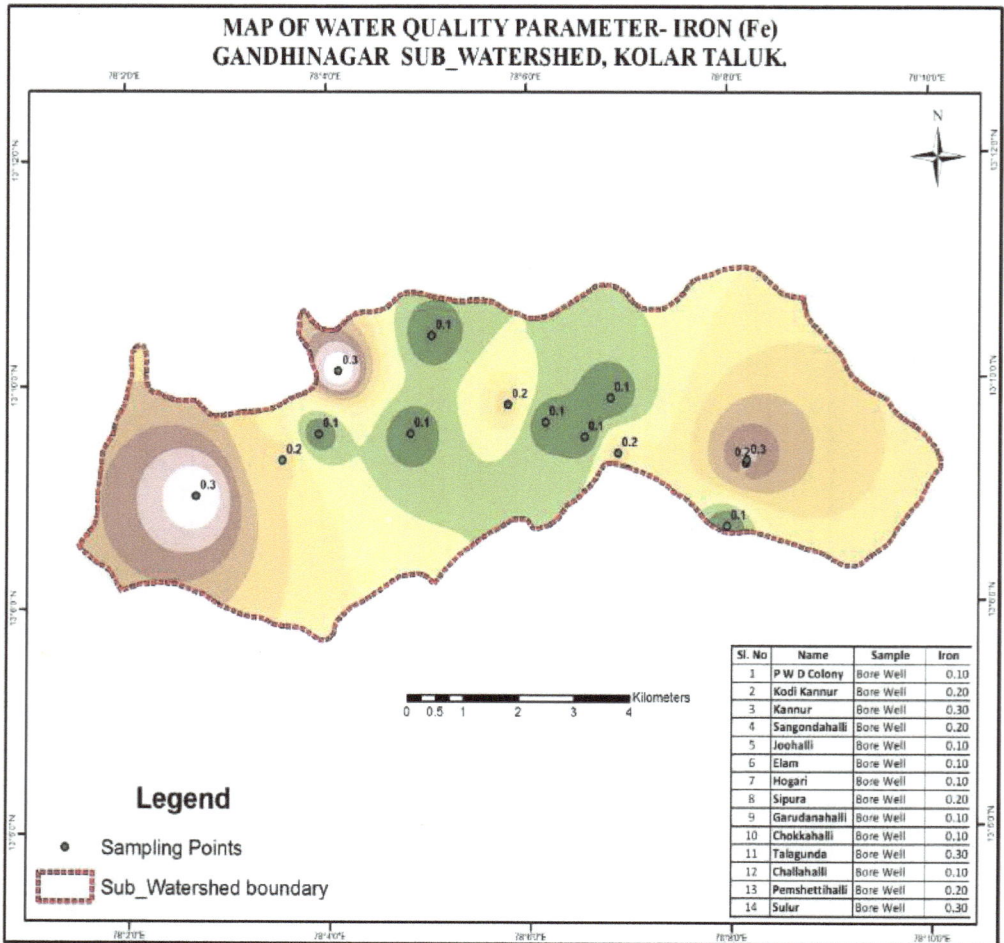

Fig. (13). Map of water quality parameter with respect to iron.

The recommendations for preventing further groundwater quality:

1. Quantifying the domestic sewage that enters the various water bodies located within the city will help plan for an adequate sewage treatment plant and minimize groundwater pollution by sewage.

2. Groundwater recharging structures are to be formed in different parts of the town and within the villages. The formation of stormwater drains results in groundwater recharging structures, extending their recharging potentials.

3. Continuous monitoring of groundwater table levels alongside quality study will minimize the probabilities of further deterioration.

Structural engineers, consultants, contractors, and the public are to be addressed about the groundwater quality not satisfying the water quality requirements as per BIS and advising them to avoid untreated groundwater.

CONCLUSION

In this study, a detailed analysis of various physicochemical parameters of samples collected from 14 bore wells revealed total hardness, calcium, magnesium, fluoride, and slightly alkaline in the Gandhinagar sub-watershed region in the Kolar district. All villages are found to be having chloride, turbidity, and iron concentration within the desirable limit. The results indicate the need to make the public, local administrators, and, therefore, the government aware of the crisis of poor groundwater quality prevailing within the area. The government must make a clear and feasible plan for identifying an efficient groundwater quality management system and its implementation. Thus, public awareness of this quality crisis and its involvement and cooperation with local administrators' actions are vital. Since the groundwater will have the leading share of water system schemes, plans for the protection of groundwater quality are required. The present status of groundwater necessitates continual monitoring, and necessary groundwater quality improvement methodologies must be implemented.

CONSENT FOR PUBLICATION

Not applicable.

CONFLICT OF INTEREST

The author declares no conflict of interest, financial or otherwise.

ACKNOWLEDGEMENT

Declared none.

REFERENCES

[1]   J. Umamaheswa, R. Anjali, S. Abinandan, S. Shanthakum, G.P. Ganapathy, and M. Kirubakara, "Assessment of Groundwater Quality Using GIS and Statistical Approaches", *Asian Journal of Earth Sciences,* vol. 8, no. 4, pp. 97-113, 2015.

[http://dx.doi.org/10.3923/ajes.2015.97.113]

[2]   K.J. Fatombi, T.A. Ahoyo, O. Nonfodji, and T. Aminou, "Physico-chemical and bacterial characteristics of groundwater and surface water quality in the Lagbe town: Treatment essays with Moringa oleifera seeds", *J. Water Resource Prot.,* vol. 4, no. 12, pp. 1001-1008, 2012. [http://dx.doi.org/10.4236/jwarp.2012.412116]

[3]   A.R.K. Kulandaivel, P.E. Kumar, V. Perumal, and P.N. Magudeswaran, "Water quality index of river cauvery at erode region, Tamilnadu, India", *Nature Environ. Pollut. Technol.,* vol. 8, pp. 343-346, 2009.

[4]   S. Senthilkumar, and T. Meenambal, "Study of groundwater quality near Sipcot industrial estate of Perundurai of Erode district, Tamilnadu", *Nat. Environ. Pollut. Technol.,* vol. 6, pp. 741-744, 2007.

[5]   R.K. Ranjan, A. Ramanathan, P. Parthasarathy, and A. Kumar, "Hydrochemical characteristics of groundwater in the plains of Phalgu River in Gaya, Bihar, India", *Arab. J. Geosci.,* vol. 6, no. 9, pp. 3257-3267, 2013. [http://dx.doi.org/10.1007/s12517-012-0599-1]

[6]   P. Ravikumar, M. Aneesul Mehmood, and R.K. Somashekar, "Water quality index to determine the surface water quality of Sankey tank and Mallathahalli lake, Bangalore urban district, Karnataka, India", *Appl. Water Sci.,* vol. 3, no. 1, pp. 247-261, 2013. [http://dx.doi.org/10.1007/s13201-013-0077-2]

[7]   M. Thiyagarajan, and R. Baskaran, "Groundwater quality in the coastal stretch between Sirkazhi and Manampandal, Tamil Nadu, India using ArcGIS Software", *Arab. J. Geosci.,* vol. 6, no. 6, pp. 1899-1911, 2013. [http://dx.doi.org/10.1007/s12517-011-0500-7]

[8]   S.K. Pandey, and S. Tiwari, "Physico-chemical analysis of ground water of selected area of Ghazipur city-A case study", *Nat. Sci.,* vol. 7, pp. 17-20, 2009.

[9]   American Public Health Association (APHA), "Standard methods for the examination of water and wastewater", In: *American Public Health Association, American Water Works Association, Water Environment Federation.* Washington, DC, 1998.

[10]   "BIS (Bureau of Indian Standards) 10500", *Indian standard for drinking water specification,* pp. 1-8, 1991.

[11]   WHO, *Guidelines for Drinking Water Quality.* 3rd ed. World Health Organization: Geneva, Switzerland, 2011, pp. 296-459.

[12]   B.P. Naveen, and T.G. Durga Madhab Mahapatra, "Physico-chemical and biological characterization of urban municipal landfill leachate", In: *Environmental Pollution* Elsevier, 2017, pp. 1-12.

[13]   CPCB, "Guidelines for water qualitymanagement", In: *Central Pollution Control Board* Parivesh Bhawan: East Arjun Nagar, Delhi, 2008.

[14]   T. Ramkumar, S. Venkatramanan, I. Anithamary, and S.M.S. Ibrahim, "Evaluation of hydrogeochemical parameters and quality assessment of the groundwater in Kottur blocks, Tiruvarur district, Tamilnadu, India", *Arab. J. Geosci.,* vol. 6, no. 1, pp. 101-108, 2013. [http://dx.doi.org/10.1007/s12517-011-0327-2]

[15]   S.A. Abbasi, "Water Quality Sampling andAnalysis", *Discovery Publishing House, NewDelhi.,* p. 51, 1998.

CHAPTER 11

Threats to Sustainability of Land Resources Due to Aridity and Climate Change in the North East Agro Climatic Zone of Tamil Nadu, South India

P. Dhanya[1,*], A. Ramachandran[1] and **K. Palanivelu[1]**

[1] *Centre for Climate Change and Adaptation Research, Anna University, Chennai, India*

Abstract: Focusing on the erstwhile Chengalpattu district, north-east agro-climatic zone of Tamil Nadu region, this research aimed to assess the changes in spatio-temporal patterns and trends of the extreme climatic events and aridity conditions during the period 1971-2000. The trend analysis of the observed climatic parameters was carried out using R software and Mann-Kendall non-parametric test. A statistically significant increasing trend was noted in the warm spell duration (wsdi) and heavy rainfall events (r_{20}mm). The results revealed that Aridity Index (AI) has significant negative trends in northeast monsoon and winter seasons, indicating dryness, and positive trends in southwest monsoon seasons, indicating wet climate. The trends in MI were found to be mostly negative during the southwest monsoon season. The results of the trend analysis in PET revealed a significant increase annually and seasonally. Overall, spatial analysis characterized the western parts as semi-arid, whereas a dry sub-humid climate prevails in the eastern parts, covering the coastal areas. As per the outcome, there may be escalations of 19.5 to 25.7% in PE in the study area. Parts of Kancheepuram, Sriperumbudur, Chengalpattu, Thirukazhikundram, Maduranthakam, and the whole of Uthiramerur blocks are going to be severely impacted due to the rise of PE. This may further trigger an escalation of aridity processes in the future and pose threats to the sustainability of land resources.

Keywords: Aridity, Climate change, Land degradation, Land resources, Northeast agro-climatic zone, Sustainability, Tamil Nadu.

INTRODUCTION

According to the United Nations Convention to Combat Desertification, about 32% of India's land is being affected by land degradation. India stands strongly committed to implementing the UNCCD goals (National Report 2010) [1, 2]. In this context, research works involving regional and local assessment of the changing patterns of aridity deserve prime attention. Aggravations in the paridity

---

* **Corresponding author P. Dhanya:** Centre for Climate Change and Adaptation Research, Anna University, Chennai, India; Tel: +91-44-22357947/7949; E-mail: dhanya.eptri@gmail.com

G. Venkatesan, S. Lakshmana Prabu and M. Rengasamy (Eds.)

conditions act as a stress multiplier in the context of global climate change [3]. The latest estimates of future climate change for India by IPCC (2013) project that there will be an increment of 5-10 days in the consecutive dry days as per emission pathway RCP 8.5 for the period 2081-2100 [4]. The rising frequency of dry spells and drought as a serious consequence of climate change was also reported in the earlier reports of the Intergovernmental Panel on Climate Change [5, 6]. Cruz *et al.* reported that the Indian subcontinent would be adversely affected by enhanced climate variability with rising temperatures and substantial reduction of summer rainfall in some parts, and associated acute water stress by the 2020s [7]. As per the predictions, the arid and semi-arid regions are going to be impacted severely by heavy land degradation, loss of biodiversity, and food insecurities [8, 9]. Water is going to be the major limiting factor for the healthy functioning of ecosystems, especially agriculture production worldwide [10, 11].

In general, several studies have been undertaken to analyze the trends in long-term precipitation and temperature, and its inter-annual, seasonal, and decadal variability at different time scales [10, 12 - 14]. Researchers have focused and extensively reported on regional aridity issues worldwide [15 - 18]. The findings from the analysis revealed that the sites which are currently at the limit with respect to available water resources, especially in semi-arid regions, are likely to be most sensitive to climate change challenges [19 - 23].

As per the UNCCD Report, 2001, dry lands are described as arid (excluding the polar and sub-polar regions), semi-arid, and dry sub-humid areas where the distribution of annual rainfall to potential evapotranspiration ratios falls within the range from 0.05 to 0.65 [9]. The arid and semi-arid regions comprise almost 40% of the world's land surfaces. The latest CMIP5 ensemble means future projections of RCP6 and RCP 8.5 emission pathways indicate a rise in the warming of about 3.3 to 4.8° C by 2080s, relative to pre-industrial times [24]. A rise in the day time extreme temperatures, hot days, and increasing dryness have also been predicted for South Asian Regions [5, 6, 25].

India has been confronted with multiple challenges, from the rapid growth of population combined with drastic land-use changes and huge demand for food production. Under this circumstance, water scarcity may stand as a serious concern for agricultural production in arid and semi-arid areas of India due to its increasing over-dependency on the summer monsoon rainfall. NATCOM reports that climate stress is coupled with its unpredictability and inefficient irrigation Dev.elopments in our country [1]. Some researchers have established that teleconnections play a vital role in deciding climate change variability. Droughts events will have a strong link with the EI Nino-Southern Oscillation (ENSO) patterns [13, 26]. As per the Indian scenario, the recent drought of 2002 and 2004 suggests the inherent vulnerability of the Indian monsoon system due to the EI Nino phenomenon [27, 28]. In this study, the observed climate variables, temperature, rainfall, and PET have been utilized for analyzing the historical trends, aridity index, and moisture index. Apart from that simulation output of the Regional Climate Model, RegCM under RCP 4.5 emission trajectory has been utilized to provide a glimpse of the future likely PE for the study area.

MATERIALS AND METHODS

Study Area

Chengalpet (erstwhile known by this name), covering the present Thiruvallur and Kancheepuram district (Fig. 1), forms part of the north-east agro-climatic zone of Tamil Nadu. It is situated between the latitudes 12° 0' 0" and 13° 40' 0" and longitude 79°0' 0" to 80° 20' 0". This area has a coastline of 115.1 km. The general slope of the study area is from northwest to southeast direction. The elevation of this area is between 100-200 meters.

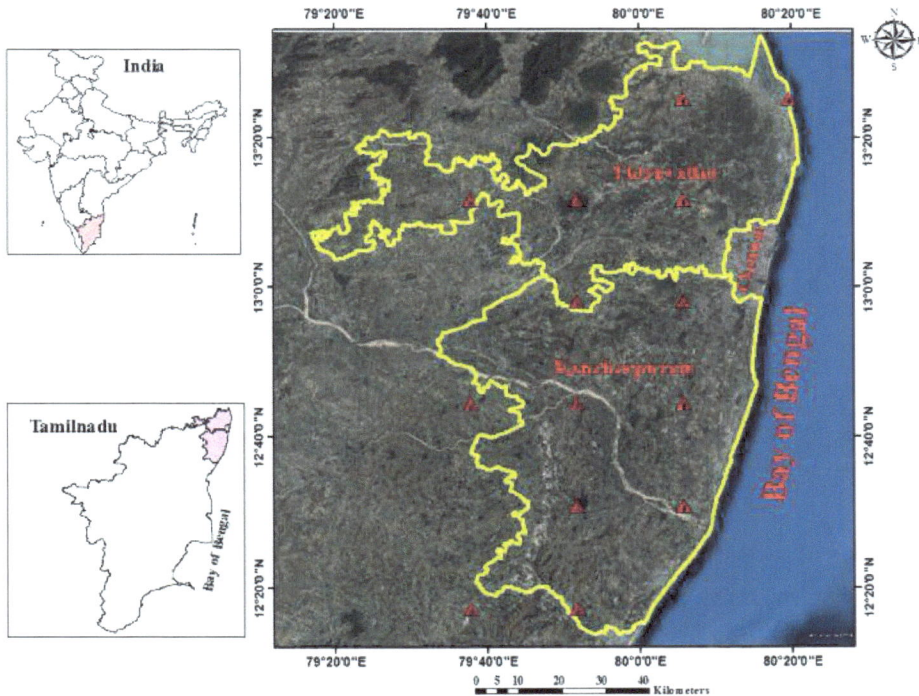

Fig. (1). The study region.

This area receives rain under the influence of both southwest and northeast monsoons. Most of the rain that occurs in the northeast monsoon season is due to cyclonic storms caused due to the depressions in the Bay of Bengal, chiefly during the October, November, and December months. The southwest monsoon rainfall is highly erratic. Palar, Cheyyar, Araniyar, and Kosasthalaiyar are important rivers. All these rivers are seasonal and carry substantial flows only during the monsoon period. The chief irrigation sources in the area are tanks, wells, tube wells, and canals. The maximum daily temperature rarely exceeds 43° C, and the minimum daily temperature seldom falls below 18° C. The annual mean rainfall of this area is 1202 mm, and potential evapotranspiration of 1750 mm. The rainfall in this region shows remarkable annual variability.

These areas were known for temple tanks once upon a time. The decline in the area under water bodies, over-extraction of groundwater, and loss of soil fertility have been noted owing to the vagaries of monsoon, and land-use changes. As per the Seasonal Crop Report 2009-2010, the net sown area covers 58.3% of the total geographic area of the district [29]. Paddy, groundnuts, sugarcane, and pulses are the major crops cultivated in this region. Tanks, wells, and tube wells are the major sources of irrigation.

MATERIALS AND METHODS

Climatic Datasets

The precipitation and potential evapotranspiration data sets of the various 10 stations pertaining to the study area have been collected from the climate research unit of Tamil Nadu Agriculture University, Coimbatore, for the period 1951 to 2008. This was taken as the input data for computing the Aridity Index as per UNEP classification and Moisture Index as per Thornthwaite and Mather [18, 30]. MI formula was further simplified by Venkataraman and Krishnan [31]. The climatic parameters of the stations, such as Chittamur, Maduranthakam, Thirukazhikundram, Kancheepuram, Wallajabad, St. Thomas Mount, RK Pet, Poondi, Villivakkam, and Gummudi Poondi, were collected for the analysis.

Aridity and Moisture Index Calculations

The aridity index, Dev.eloped by the United Nations Environmental Program (UNEP), has been utilized to understand the observed aridity status of the study area.

$$\text{Aridity Index (AI)} = \text{Precipitation/PET} \tag{1}$$

Table 1. Categories of aridity index.

Zone Aridity Index values
Hyper-arid < 0.05
Arid 0.05 - 0.2
Semi-arid 0.2 – 0.5
Dry Humid 0.5 – 0.65
Humid >0.65

The rainfall and PE values for each station were put into the formula of moisture index given by Thornthwaite and Mather and simplified by Venkataraman and Krishnan as shown below:

$$\text{Moisture Index (MI)} = [\text{P-PE}]/\,\text{PE} * 100 \qquad (2)$$

Where MI is the moisture index, P is the average annual rainfall, and PE is the average annual potential evapotranspiration. Further, the climate prevailing in each station during the study period was assessed as following:

Table 2. Categories of moisture index.

Value of MI Climatic Zone
< –66.7 Arid
–66.6 to –33.3 Semi-arid
–33.3 to + 20 Sub-humid
+20.1 to + 99.9 Moist sub-humid
100 or more Humid Per-humid

Table 3. Categories of extremity indices.

Indices	Codes	Description	Unit
Warm spell duration index	WSDI	The annual number of days with at least 6 consecutive days when Tmax>90th percentile	Days
Very heavy precipitation days	R20 mm	Annual count when precipitation is more than 20 mm	Days

EXTREMITY ANALYSIS

The user-friendly R-based software module RClimDex was used for analysis (http://cccma.seos.uvic.ca/ETCCDI). To gain a detailed understanding of climate extremes, the ETCCDI defined a core set of descriptive indices based on characteristics like frequency, amplitude, and persistence [20, 32, 33]. The indices that have been included in this paper are warm spell duration (wsdi) and R20mm rainfall.

Trend Analysis

Trends in any data series can be identified through parametric or non-parametric methods. The main advantages of non-parametric methods are they are effective even if the distribution is not normally distributed and have several outliers. Mann-Kendall is a non-parametric trend test basically to detect the trends in the data series, as described by Sneyer (1990), was applied in order to detect trends. This test has been widely used to detect trends in hydrological time series data, and the methodologies have been reported in detail by several researchers [11, 17, 34 - 36]. A positive value of *tau* indicates an upward trend, and a negative value of *tau* indicates a downward trend. The direction and magnitude of the trends have been evaluated using Sen's Slope estimator.

Future Projected Potential Evaporation for the Period 2070-2100

Data on climate change projections were processed using regional climate model-RegCM 4.4 rc22, with HadGEM 2-ES lateral boundary conditions. It is an earth system model which was run at a regional level with a resolution of 25 km * 25 km, especially for Tamil Nadu. For simulating the plausible future climate, the latest available emission trajectory, Representative Concentration Pathway 4.5 (RCP4.5) proposed by the AR5 report of IPCC, has been used. The simulation was carried out from 1970 to 2100 continuously. Future likely Potential evaporation values pertaining to the Chengalpet area have been utilized for projections. Model evaluation was performed to ensure that the RegCM model output matched the corresponding statistics of the observed data for the baseline period 1970-2000. ArcGIS 10.1 version has been used to explore and analyze the output of the model. The relative change of PE for the end of the 21st century was calculated based on the reference period 1970-2000.

RESULTS AND DISCUSSION

The Trend Analysis of Observed Aridity

The results of the MK test and Sen's slope estimator reveal that there is a statistically significant downward trend in aridity values during the northeast (OND) and winter season (JF) for all the stations over a period of 57 years (Table 1). Fig. (2) Sen's slope estimated the highest negative trend value of -0.02 for AI during the northeast monsoon season. Aridity conditions worsened during the northeast monsoon season as a high negative value of *'tau'* was observed, indicating a drying trend. Considering the entire data series for the period 1951-2008, it has been detected that the majority of the stations exhibited negative values during these two seasons. However, the aridity index exhibited significant positive trends during the southwest season (JJAS). But there was no significant trend detected for the annual mean aridity index. RK Pet station recorded the lowest annual mean AI value of 0.464, and maintained consistency across all the seasons. AI values showed a high standard Dev.iation, mainly during winter. There was no significant trend noticed for the summer (MAM) season for any of the stations except for RK Pet. The lowest recorded AI value among all the stations was 0.003, which was mainly in the southwest monsoon season.

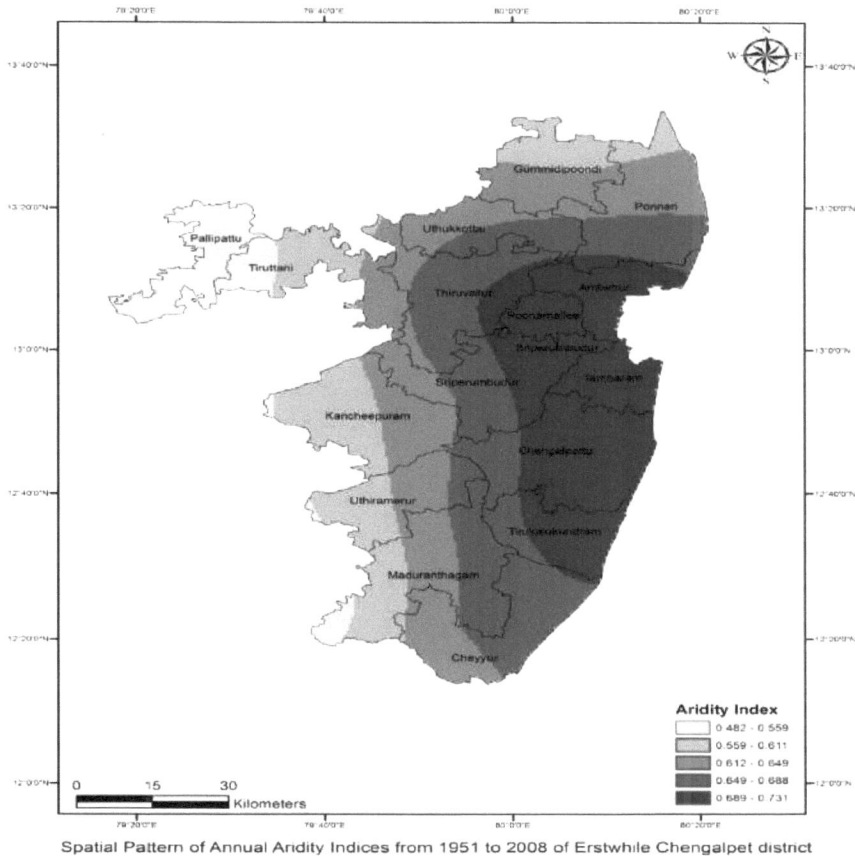

Spatial Pattern of Annual Aridity Indices from 1951 to 2008 of Erstwhile Chengalpet district

Fig. (2). Spatial pattern cf average annual aridity index from 1951 to 2008.

The linear trends show the temporal pattern of mean annual aridity during the period 1951-2008. AI during the year period 1958-59, 1968-69, 1974-75, 1983-84, 2002-2003 exhibited sharp declining trends and had reached between 0.2 to 0.3 in a majority of the stations. This indicated that during this period, this area was hit by severe drought conditions.

Trend Analysis of Observed Moisture Status

Similar to AI, MI also showed a statistically significant downward trend, especially in the northeast and winter seasons during the study period 1951-2008 (Table **2** and Fig. **3**). The negative trend was maintained in all the stations. Sen's slope detected the highest negative slope value of -2.46, during the northeast season. The mean value of RK Pet stations seemed to be consistent as the MI values ranged from -41.7 to -55.4 during all the seasons. However, all the stations exhibited an upward trend during the southwest monsoon season as the value of *'tau'* was positive. MI values showed high standard Dev.iation mainly in the winter season and summer seasons in the majority of the stations, especially Thirukazhikundram, St. Thomas Mount, Chittamur and Villivakkam. Nevertheless, there was no significant trend detected for annual mean MI except for Madhuranthakam and Gummudipundi (Table **4**). These two stations showed significant upward trends with a sen's slope values of 0.114 and 0.292, respectively.

Table 4. Results of the MK test and Sen's slope test on the aridity index for the 10 stations in Chengalpet areas.

AI_MK TEST								
Location	Season	Sen's Slope	p	Mean	Max	Min	Std. Dev.	Tau
Chittamur	JF*	-0.014	0.001	0.776	4.029	0.039	0.829	-0.313
	MAM	0.007	0.051	0.743	3.027	0.02	0.739	0.178
	JJAS*	0.017	0.003	0.619	2.752	0.003	0.637	0.272
	OND*	-0.024	0.0001	0.498	1.785	0	0.329	-0.684
Madhuranthagam	JF*	-0.015	0.001	0.767	3.474	0.036	0.739	-0.298
	MAM	0.009	0.016	0.733	2.91	0	0.728	0.221
	JJAS*	0.017	0.003	0.631	2.713	0.003	0.678	0.269
	OND*	-0.023	0.0001	0.513	1.814	0	0.0544	-0.675
Thirukazhukundaram	JF*	-0.017	0	0.831	3.592	0.037	0.818	-0.321
	MAM	0.009	0.044	0.843	3.506	0	0.863	0.184
	JJAS*	0.021	0.001	0.71	3.092	0.003	0.778	0.302
	OND*	-0.025	0.0001	0.534	1.929	0	0.563	-0.66
Kanchee	JF*	-0.012	0.008	0.692	2.829	0.012	0.58	-0.242
	MAM*	0.01	0.003	0.626	2.233	0	0.558	0.276
	JJAS*	0.013	0.008	0.544	2.246	0.003	0.533	0.242
	OND*	-0.024	0.0001	0.503	1.609	0	0.494	-0.684

(Table 4) cont.....

AI_MK TEST								
Location	**Season**	**Sen's Slope**	**p**	**Mean**	**Max**	**Min**	**Std. Dev.**	**Tau**
Wallajabad	JF*	-0.017	0.001	0.76	3.361	0.009	0.689	-0.308
	MAM	0.004	0.013	0.727	2.933	0	0.689	0.226
	JJAS*	0.018	0.001	0.619	2.691	0.002	0.648	0.291
	OND*	-0.024	0.001	0.519	1.854	0	0.515	-0.678
St. Thomas mount	JF*	-0.017	0.001	0.821	3.347	0.025	0.781	-0.308
	MAM	0.01	0.028	0.85	3.52	0	0.836	0.201
	JJAS*	0.022	0.001	0.714	3.326	0.003	0.794	0.31
	OND*	-0.025	0.0001	0.542	1.884	0	0.546	-0.65
Rk pet	JF*	-0.011	0.004	0.584	1.817	0.041	0.395	-0.266
	MAM*	0.011	0.001	0.524	1.636	0	0.449	0.315
	JJAS	0.009	0.015	0.45	1.749	0.003	0.426	0.223
	OND*	-0.022	0.0001	0.461	1.807	0	0.449	-0.716
Poondi	JF*	-0.017	0.001	0.787	3.23	0.003	0.698	-0.296
	MAM	0.011	0.01	0.778	8.036	0	0.709	0.234
	JJAS*	0.02	0.001	0.663	2.804	0.002	0.69	0.294
	OND*	-0.025	0.0001	0.533	1.846	0	0.518	-0.673
Villivakkam	JF*	-0.02	0	0.833	3.424	0.013	0.779	-0.331
	MAM	0.01	0.042	0.87	3.319	0	0.81	0.185
	JJAS*	0.024	0	0.738	3.502	0.001	0.841	0.328
	OND*	-0.024	0.0001	0.542	1.933	0	0.54	-0.654
Gummudipoondi	JF*	-0.016	0	0.705	2.852	0.005	0.659	-0.338
	MAM	0.008	0.084	0.709	2.465	0	0.608	0.158
	JJAS*	0.02	0	0.644	3.127	0.002	0.72	0.33
	OND*	-0.018	0.0001	0.449	1.582	0	0.43	-0.635

*denotes trends are significant at 0.01 level.

Table 5. Results of the MK test and Sen's slope test on the moisture index for the 10 stations in Chengalpet areas.

AI_MK TEST								
Location	**Season**	**Sen's Slope**	**p**	**Mean**	**Max**	**Min**	**Std. Dev.**	**Tau**
Chittamur	JF*	-1.441	0.001	-22.41	302.94	-96.18	82.92	-0.313
	MAM	0.69	0.051	-25.74	202.74	-99.83	73.88	0.18
	JJAS*	1.721	0.003	-38.06	175.17	-99.67	66.97	0.27
	OND*	-2.357	0.0001	-50.25	78.493	-99.96	52.91	-0.69
Madhuranthagam	JF*	-1.498	0.001	-23.31	247.35	-96.37	73.883	-0.29
	MAM	0.874	0.016	-26.72	191.71	100	72.792	0.22
	JJAS*	1.675	0.003	-36.86	171.29	-99.72	67.78	0.27
	OND*	-2.349	0.0001	-48.84	81.41	-99.98	54.417	-0.68

(Table 5) cont.....

AI_MK TEST								
Location	Season	Sen's Slope	p	Mean	Max	Min	Std. Dev.	Tau
Thirukazhukundaram	JF*	-1.667	0	-16.865	259.178	-96.29	81.844	-0.32
	MAM	0.858	0.044	-15.682	250.625	-100	86.349	0.18
	JJAS*	2.137	0.001	-28.965	209.202	-99.74	77.774	0.31
	OND*	-2.459	0.0001	-46.571	92.8	-99.95	56.3	-0.66
Kanchee	JF*	-1.159	0.008	-30.848	182.919	-98.77	57.98	-0.24
	MAM*	1.016	0.003	-37.389	123.314	-100	55.77	0.28
	JJAS*	1.301	0.008	-45.602	124.602	-99.79	53.31	0.24
	OND	-2.429	0.0001	-49.752	60.888	-100	49.38	-0.69
Wallajabad	JF*	-1.655	0.001	-24.02	236.064	-99.15	68.91	-0.31
	MAM	0.946	0.013	-27.282	193.343	-100	68.91	0.23
	JJAS*	1.779	0.001	-38.098	169.126	-99.79	64.78	0.29
	OND*	-2.417	0.0001	-48.889	85.351	-100	51.49	-0.69
St. Thomas mount	JF*	-1.746	0.001	-17.947	234.708	-97.47	78.143	-0.308
	MAM	0.978	0.028	-15.026	251.96	-100	83.643	0.201
	JJAS*	2.25	0.001	-28.627	232.56	-99.73	79.39	0.31
	OND*	-2.472	0.0001	-45.768	88.392	-100	54.58	-0.65
Rk pet	JF	-0.575	0.154	-41.749	146.294	-100	48.89	-0.13
	MAM*	1.028	0.0001	-47.433	63.59	-100	44.8	0.311
	JJAS	0.919	0.020	-54.742	74.872	-99.69	42.53	0.212
	OND*	-2.049		-55.433	80.569	-100	43.23	-0.692
Poondi	JF*	-1.666	0.001	-21.31	222.99	-99.75	69.835	-0.296
	MAM	1.114	0.01	-22.18	203.59	-100	70.913	0.234
	JJAS*	2.011	0.001	-33.724	180.36	-99.83	68.96	0.294
	OND*	-2.463	0.0001	-46.648	84.64	-100	51.83	-0.667
Villivakkam	JF*	-2.01	0	-16.704	242.395	-98.65	77.92	-0.34
	MAM	1.018	0.042	-13.04	231.863	-100	80.97	0.19
	JJAS*	2.355	0	-26.224	250.176	-99.88	84.15	0.33
	OND*	-2.438	0.0001	-45.662	93.26	-99.98	54.02	-0.66
Gummudipoondi	JF*	-1.633	0	-29.545	188.163	-99.5	65.364	-0.338
	MAM	0.762	0.084	-29.139	146.49	-100	60.761	0.158
	JJAS*	2.011	0	-35.596	212.727	-99.76	71.991	0.33
	OND*	-1.848	0.0001	-54.91	58.175	-99.97	42.981	-0.63

*denotes trends are significant at 0.01 level.

Time series of the linear trend in the distribution of mean annual moisture index during the period 1951-2008 (Table **5**). MI showed sharp declines during the year period 1951-52, 1958-59, 1968-69, 1973-74, 1983-84, 2002-2003 exhibited sharp declining trends and ranged between -40 to -60 in the majority of the stations. This indicates that the study area comes mainly under a semi-arid climate.

Extremities

The warm spell duration index (WSDI) averaged over the study area (Fig. **4**) showed a significant increasing trend with a rate of 3.07 days per decade since 1971(p< 0.01). The WSDI indices in 1981, 1985, 1999 and 1994 showed 37, 35, 30 and 29 number of days with at least 6 consecutive days when Tmax>90th percentile annually.

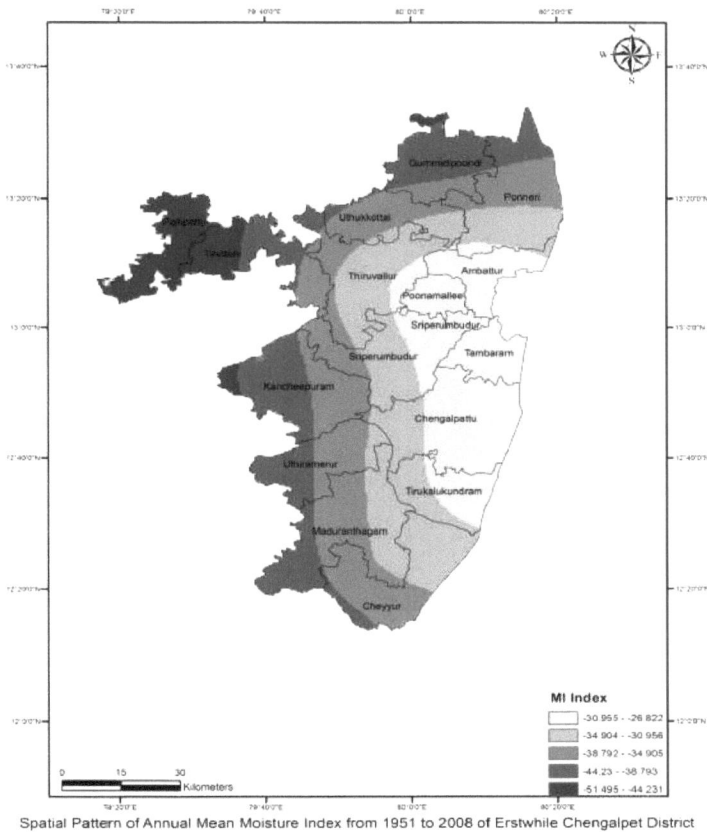

Spatial Pattern of Annual Mean Moisture Index from 1951 to 2008 of Erstwhile Chengalpet District

Fig. (3). Spatial pattern of moisture index from 1951 to 2008.

A number of very heavy precipitation days (R20 mm) averaged over the study area (Fig. **5**) showed increasing trends during the 1970s to 2000, with rates of 0.59 days per decade, respectively (p< 0.05). During the period 1970 to 2000, the years

1977, 1986, 1988, 1992, 1996 and 1997, R20 mm rainfall was received for 8-7 days.

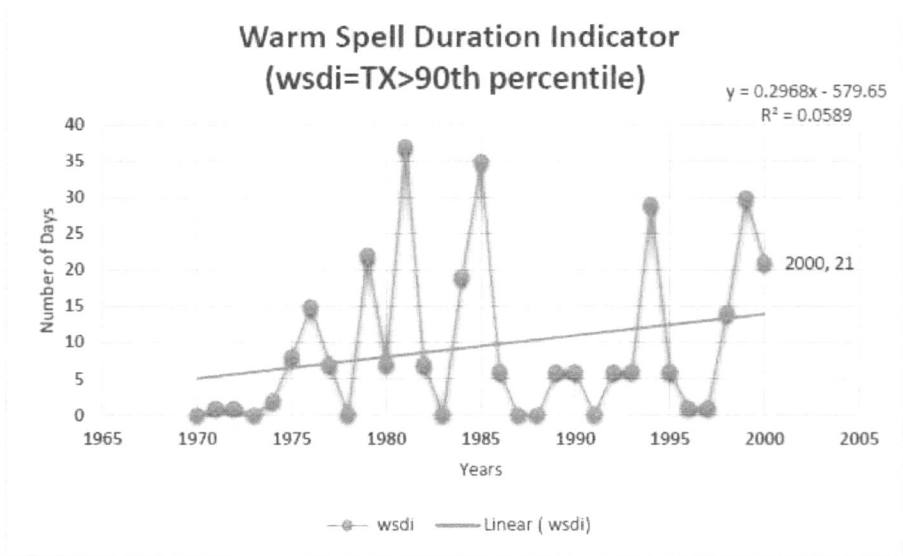

Fig. (4). Shows warm spell duration –annual temperature above 90th percentile.

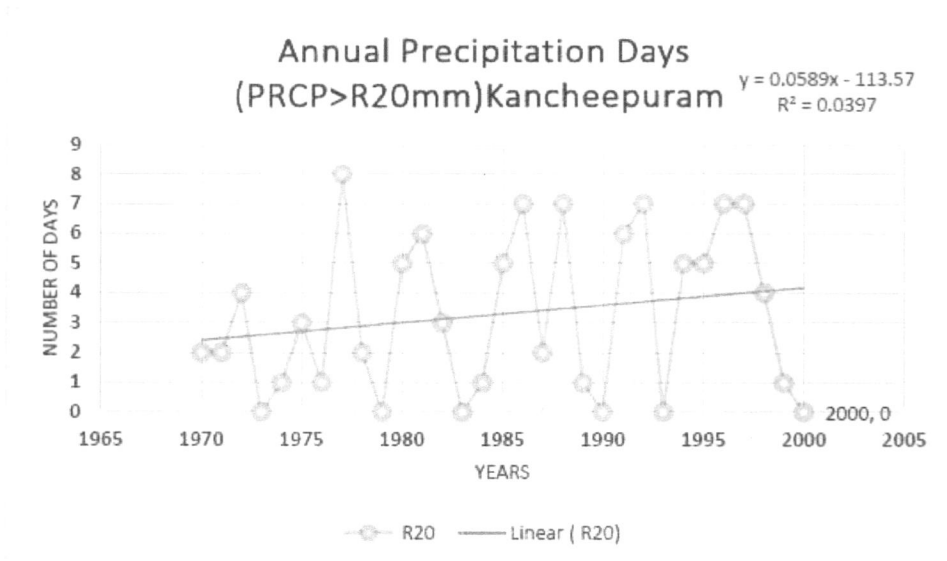

Fig. (5). Shows annual precipitation climate extremities over 20mm and above during the reference period 1970-2000.

Observed Trends in Potential Evapotranspiration

The mean annual Potential Evapotranspiration recorded during the study period 1951-2008 was approximately 1750 mm/year for a majority of the stations. Contrary to AI & MI and PET showed a statistically significant downward trend during southwest monsoon for all the stations except for Wallajabad during the study period 1951-2008 (Table **6** and Fig. **6**). The highest negative slope value of -5.45 mm/season was detected through Sen's slope estimator for Poondi station. On the other hand, an upward trend was noted in all the stations for the winter and northeast seasons. The highest recorded positive slope value was 3.57mm for the Poondi station and 3.55 mm for the RK pet stations. There was a slightly increasing trend noted for annual mean PET values, especially for Chittamur and Maduranthakam, however not statistically significant (Table **6**). There was a high standard Dev.iation noticed during southwest monsoon seasons in almost all the stations. RK Pet recorded the highest maximum PET of 1839mm/year among all the stations during the entire study period.

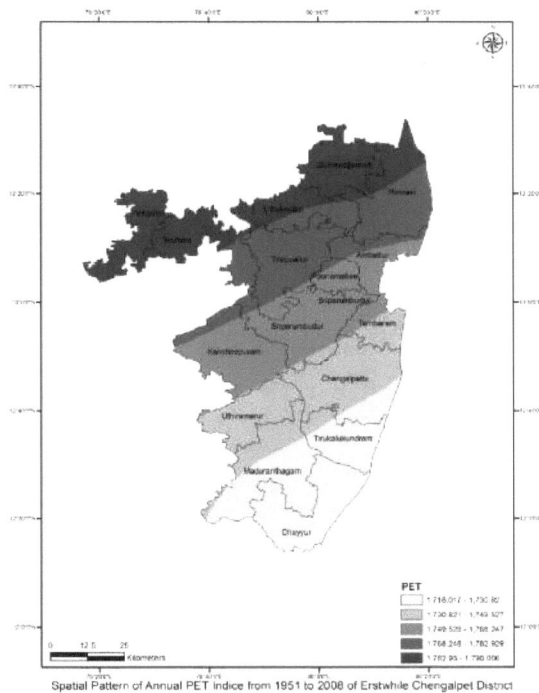

Spatial Pattern of Annual PET indice from 1951 to 2008 of Erstwhile Chengalpet District

Fig. (6). Spatial pattern of annual average PET from 1951 to 2008.

The linear trend of the time series distribution of PET during the period 1951-2008. PET hit its peak during the year period 1964-65, 1983-84, 1998-199, 2002-2003 in the majority of the stations. A sharp decline in PET was noted in the year period 1954-55.This indicates that the increment in PET corresponds well with the AI and MI values over many years.

Table 6. Results of the MK test and Sen's slope test on the potential eapotranspiration for the 10 stations in Chengalpet areas.

PET_MKTEST								
Location	Season	Sen's slope	*P*	Mean	Maximum	Minimun	Std. Dev.	Tau
Chittamur	JF*	3.4	0.0001	287.3	379.6	209.7	56.48	0.75
	MAM	0.97	0.22	413.2	536.3	316.8	62.98	0.12
	JJAS*	5.16	0.0001	585.2	740.4	457.01	94.83	-0.65
	OND*	2.36	0.001	459.9	561.3	348.5	65.9	0.29
Madhuranthagam	JF*	3.39	0.0001	285.79	374.87	209.21	55.67	0.76
	MAM	0.91	0.234	411.69	525.63	316.54	62.04	0.11
	JJAS*	-5.08	0.0001	582.35	733.3	457.45	92.41	-0.64
	OND*	2.58	0	457.71	557.1	347.15	65.21	0.38
Thirukazhukundaram	JF*	3.35	0.0001	284.21	374.31	207.85	55.21	0.74
	MAM	0.99	0.184	409.19	529.39	314.05	61.49	0.13
	JJAS*	-5.03	0.0001	577.66	728.85	454.94	91.74	-0.65
	OND*	2.22	0.002	454.29	551.44	345.79	64.09	0.29
Kanchee	JF*	3.48	0.0001	289.41	380.57	213.22	56.88	0.77
	MAM	0.69	0.421	417.97	536.25	321.35	63.59	0.08
	JJAS*	-5.29	0.0001	594.02	747.14	463.75	96.04	-0.63
	OND*	3.18	0.0001	465.59	570.54	351.5	67.88	0.37
Wallajabad	JF*	-5.25	0.0001	589.755	744.55	459.84	95.64	-0.62
	MAM*	3.46	0.0001	415.675	380.69	210.99	56.98	0.75
	JJAS	0.82	0.312	415.675	532.83	318.86	63.549	0.09
	OND*	2.74	0	463.111	566.65	349.85	67.04	3
	JF*	3.43	0.0001	287.27	379.63	209.65	56.49	0.75
	MAM	0.98	0.223	413.25	536.26	316.75	62.98	0.112

(Table 6) cont.....

St. Thomas Mount	JJAS*	-5.17	0.0001	585.17	740.41	457.01	94.83	-0.647
	OND*	2.362	0.001	459.89	561.25	348.43	65.89	0.291
	JF*	3.43	0.0001	292.02	386.91	217.67	57.04	0.772
	MAM	0.29	0.746	423.21	550.05	327.3	64.05	0.03
RKPET	JJAS*	-5.35	0.0001	603.68	755.45	470.15	98.32	-0.603
	OND*	3.553	0.0001	471.08	580.39	354.23	70.67	0
	JF*	3.573	0.0001	291.656	387.51	212.48	58.11	0.774
	MAM	0.662	0.39	419.756	539.26	322.84	64.63	0.079
Poondi	JJAS*	-5.449	0.0001	596.867	754.82	461.96	98.35	-0.64
	OND*	3.052	0	468.275	576.45	351.33	68.7	0.351
	JF*	2.49	0.0001	465.352	571.05	352.03	67.073	0.29
	MAM*	3.48	0.0001	290.536	385.57	212.07	57.564	0.75
Villivakkam	JJAS	0.95	0.198	417.16	541.73	320.94	63.835	0.12
	OND*	-5.29	0.0001	592.144	751	459.49	96.824	-0.64
	JF*	3.599	0.0001	307.35	390.18	219.24	60.41	0.84
Gummudipoondi	MAM*	2.587	0.0001	417.35	558.72	328.48	59.75	0.39
	JJAS*	-6.018	0.0001	579.18	747.04	440.59	104.96	-0.81
	OND	0.503	0.332	483.37	581.85	376.33	54.636	0.09

Spatial Distribution of Future Likely Potential Evaporation for the Period 2070-2100

Fig. (7) shows how predicted changes in PE are increasing towards the end century with respect to the baseline period 1970-2000. The maximum and minimum increase projected for the study area are in the range of 25.7 and 19.5%, respectively. Parts of Kancheepuram, Sriperumbudur, Chengalpattu, Thirukazhikundram, Maduranthakam and the whole of Uthiramerur blocks are going to be severely impacted due to the rise in PE.

DISCUSSION

Land degradation causes serious negative implications for biodiversity loss, the most critical global environmental threat. More than a third of the global species are facing extinction, and an estimated 60% of the earth's ecosystems have been degraded in the last 50 years, with consequences for the services that depend on them [37]. This region can be placed under the tropical semi-arid to dry sub-humid climate, as the coastal stretch comes under the latter and interior parts come under the former category. Previous studies reported by Dhanya *et al*. [38], for the same area show an increase in projected mean annual maximum and maximum temperature of about 3.3°C over the study region during 2070-2100,with reference to the baseline year, 1970-2000. The interior western parts of the districts may be more impacted by the increase in temperature than the coastal area. The projections showed slight decreasing trends for annual mean rainfall with a range of 700-900 mm/year over the western and interior north-western parts of the district during the period 2070-2100.

Fig. (7). Spatial pattern of projected PE over Chengalpet area during the period 2070-2100.

Rising temperature plays a major role in accelerating hydrological processes to a broad spatial extent and causing severe threats to the fertility of land resources. It leads to higher rates of evapotranspiration. This could be the major driver for increasing PE in the future. The rise in temperature will potentially increase the PE from the soil through increasing atmospheric demand. Even though this will not be the sole reason for the future likely increment in PE, there may be other forcing variables, such as wind speed, humidity, solar radiation, cloud cover, soil moisture *etc.*, with the atmospheric water demand, as they surely interact in a complex way. Their interactions may not exist in a straight line way. As the results articulate, there could be an escalation of aridity processes as the PE is going to rise in the range of 19.5 to 25.7% by the end century, as per the RCM projections. These findings are in accordance with the finding of Huntington (2006), which shows there could be accelerating extremes in the hydrological process under the recent and future climate change scenario.

Climate change is a key driver of soil degradation, especially in arid and semi-arid regions [22, 29, 36, 39 - 41]. Selvaraju (2003) has clearly demonstrated the linkage between ENSO and Indian food grain production [29]. An increase in drought processes is predicted for the Indian region under SRES, AIB scenario [39, 42]. Drying land areas may lead to severe land degradation and induce biodiversity and vegetative loss, the decline in agriculture production and depletion of surface and groundwater resources.

Aridity is a phenomenon that contributes to the increasing fragility of the systems impacted by multiplying the negative environmental, economic and social effects. One of the key ecosystems and economic sectors which is inherently sensitive to aridity is agriculture. Arid conditions craft this sector into the most vulnerable to the risk of climate variability and change [14, 41, 43]. However, researchers have found that locally available solutions and indigenous practices, along with planned and anticipatory adaptations, have enormous potential to combat these challenges and bring about many new opportunities [19, 24, 36, 44, 45].

CONCLUSION

This study has illustrated the results of the trend analysis of monthly, seasonal and annual aridity, moisture index and PET and also future projections of PE were performed. The results revealed statistically significant negative trends in northeast monsoon and winter seasons, indicating drying trends and positive trends during Southwest seasons, indicating wetting trends for AI and MI values over the study period. Contrary to this, PET was showing a strong significant upward trend during the northeast monsoon and winter seasons and declining trends during the Southwest monsoon seasons. The time series analysis showed sharp declines in AI AND MI values during the year period 1951-52,1958-59, 1968-69,1973-74,1983-84,2002-2003. The RCM has projected an increase in PE in the range of 25.7 and 19.5%, respectively, for the end of the 21st century over the study area. This study has revealed that PET has a high bearing on the aridity or moisture status of this region. More detailed analysis, including other forcing variables, can shed light on the causative factors for the trend patterns in Aridity and moisture availability. This study was fruitful in characterizing the study area climatically. With this knowledge and under the present situations, policy making should incorporate the spatial variation in the aridification process at various spatial scales with a short, medium and long-term perspective. Dev.eloping countries should enhance their knowledge base on climate change by focusing on a better understanding of the science of climate change and its impact, adaptation and mitigation. International funding organizations should provide support to the research activities in climate change areas in Dev.eloping countries, especially least Dev.eloping countries, to bridge the gap of participation as well as enhance capacity building.

CONSENT FOR PUBLICATION

Not applicable.

CONFLICT OF INTEREST

The author declares no conflict of interest, financial or otherwise.

ACKNOWLEDGEMENTS

The first author gratefully acknowledges the Anna Centenary Research Fellowship Grant of the state government of Tamil Nadu (under the department of civil engineering, Anna University) for providing the financial support for carrying out this research work. We acknowledge the modeling groups of the Centre for Climate Change and Adaptation Research for making their climate model simulation outputs available for analysis.

REFERENCES

[1]   NATCOM, *India's Initial National communication to the UNFCC Ministry of Environment and forest of Govt of India.*, 2004.

[2]   A.K. Mishra, and V.P. Singh, "Drought modeling – A review", *J. Hydrol. (Amst.)*, vol. 403, no. 1-2, pp. 157-175, 2011.
        [http://dx.doi.org/10.1016/j.jhydrol.2011.03.049]

[3]   "Global Warming of 1.5°C", In: *An IPCC Special Report on the impacts of global warming of 1.5°C above pre-industrial levels and related global greenhouse gas emission pathways, in the context of strengthening the global response to the threat of climate change, sustainable Dev.elopment, and efforts to eradicate poverty.*, V. Masson-Delmotte, P. Zhai, H.-O. Pörtner, D. Roberts, J. Skea, P.R. Shukla, A. Pirani, W. Moufouma-Okia, C. Péan, R. Pidcock, S. Connors, J.B.R. Matthews, Y. Chen, X. Zhou, M.I. Gomis, E. Lonnoy, T. Maycock, M. Tignor, T. Waterfield , Eds., In Press, 2018.

[4]   IPCC,   http://ipcc-wg2.gov/AR5/images/uploads/WGIIAR5-Chap4_FGDall.pdf   FINALDRAFT, Chapter 4. Terrestrial and Inland Water Systems.

[5]   IPCC 2014 Accessed on 15-10-2014,

[6]   IPCC,   http://ipcc-wg2.gov/AR5/images/uploads/WGIIAR5-Chap19_FGDall.pdf,   FINALDRAFT Chapter 19. Emergent Risks and Key Vulnerabilities.

[7]   "R.V. Cruz, H. Harasawa, M. Lal, S. Wu, Y. Anokhin, B. Punsalmaa, Y. Honda, M. Jafari, C. Li and N. Huu Ninh. Asia Climate Change 2007: Impacts, Adaptation and Vulnerability", In: *Contribution of Working Group II to the Fourth Assessment Report of the Intergovernmental Panel on Climate Change.* Cambridge University Press: Cambridge, 2007, pp. 469-506.

[8]   B. Lobell, "Crop Yield Gaps: Their Importance, Magnitudes, and Causes", *Annual Review of Environment and Resources.,* vol. 34, pp. 179-204, 2005.

[9]   UNCCD report accessed 2014, http://www.unccd.int/ActionProgrammes/india-eng2001.pdf

[10]   A. Lu, Y. He, Z. Zhang, H. Pang, and J. Gu, "Regional structure of global warming across China during the twentieth century", *Clim. Res.,* vol. 27, pp. 189-195, 2004.
[http://dx.doi.org/10.3354/cr027189]

[11]   Y. Luo, S. Liu, S. Fu, J. Liu, G. Wang, and G. Zhou, "Trends of Precipitation in Beijing River Basin, Guangdong province. China", *Hydro Proc.,* vol. 22, pp. 2377-2386, 2005.

[12]   ElNesr, M.M, "Temperature Trends and Distribution in the Arabian Peninsula", *Am. J. Environ. Sci.,* vol. 6, no. 2, pp. 191-203, 2010.
[http://dx.doi.org/10.3844/ajessp.2010.191.203]

[13]   S. Gadgi, P.N. Vinayachandran, and P.A. Francis, "Droughts of the Indian summer monsoon: role of clouds over the Indian ocean", *Curr. Sci.,* vol. 85, no. 12, pp. 1713-1719, 2003.

[14]   R. Tomozeiu, V. Pavan, C. Cacciamani, and M. Amici, "Observed temperature changes in Emilia-Romagna: mean values and extremes", *Clim. Res.,* vol. 31, pp. 217-225, 2006.
[http://dx.doi.org/10.3354/cr031217]

[15]   S. Golian, B. Saghafian, S. Sheshangosht, and H. Ghalkhani, "Comparison of classification and clustering methods in spatial rainfall pattern recognition at Northern Iran", *Theor. Appl. Climatol.,* vol. 102, no. 3-4, pp. 319-329, 2010.
[http://dx.doi.org/10.1007/s00704-010-0267-x]

[16]   M.R. Kousari, M.R. Ekhtesasi, M. Tazeh, M.A. Saremi Naeini, and M.A. Asadi Zarch, "An investigation of the Iranian climatic changes by considering the precipitation, temperature, and relative humidity parameters", *Theor. Appl. Climatol.,* vol. 103, no. 3-4, pp. 321-335, 2011.
[http://dx.doi.org/10.1007/s00704-010-0304-9]

[17]   B. Shifteh Some'e, A. Ezani, and H. Tabari, ShiftehSome'eB, "Spatiotemporal trends of aridity index in arid and semi-arid regions of Iran", *Theor. Appl. Climatol.,* vol. 111, no. 1-2, pp. 149-160, 2013.
[http://dx.doi.org/10.1007/s00704-012-0650-x]

[18]   M. Turkes, Spatial and temporal variations in precipitation and aridity index series of Turkey.*Mediterranean climate: variability and trends.,* H-J. Bolle, Ed., Springer: Berlin, 2003, pp. PP181-PP213.
[http://dx.doi.org/10.1007/978-3-642-55657-9_11]

[19]   V .K. Arora, "The use of aridity index to assess climate change effects on annual runoff", *Journal of Hydrology.,* vol. 265, no. 1-4, pp. 164-177, 2002.
[http://dx.doi.org/10.1016/S0022-1694(02)00101-4]

[20]   P. Frich, "L.V, Alexander, P, Della-Marta, B. Gleason, M. Haylock Klein Tank, A. M. G. and T. Peterson, "Observed coherent changes in climatic extremes during the second half of the twentieth century"", *Clim. Res.,* vol. 19, no. 3, pp. 193-212, 2016.http://www.intres.com/articles/cr2002/19/c019p193.pdf

[21]   J. Fuhrer, "Agroecosystem responses to combinations of elevated CO_2, ozone, and global climate change", *Agric. Ecosyst. Environ.,* vol. 97, no. 1-3, pp. 1-20, 2003.
[http://dx.doi.org/10.1016/S0167-8809(03)00125-7]

[22]   M.M. Verstraete, AB Brink, R.J Scholes, M Benistopn, and M. Stafford Smith, "Climate change and desertification where do we stand where should we go?", *Global Planet Change.,* vol. 64, no. 3-4, pp. 105-110, 2008.
[http://dx.doi.org/10.1016/j.gloplacha.2008.09.003]

[23]   S.M. Vicente-Serrano, J.M. Cuadrat-Prats, and A. Romo, "Aridity influence on vegetation patterns in the middle Ebro Valley (Spain): Evaluation by means of AVHRR images and climate interpolation

techniques", *J. Arid Environ.,* vol. 66, no. 2, pp. 353-375, 2006.
[http://dx.doi.org/10.1016/j.jaridenv.2005.10.021]

[24]   R.K. Chaturvedi, J. Joshi, M. Jayaraman, G. Bala, and N.H. Ravindranath, "Multi model climate change projections for India under representative concentration pathways", *Curr. Sci.,* vol. 103, no. 7, pp. 1-12, 2012.

[25]   "IPCC AR5 working group1: CC 2013", *The physical science basis. Regional Climate Change: Findings of IPCC AR5 WGI, chapter 14.,* 2013.

[26]   P. Guhathakurta, O.P. Sreejith, and P.A. Menon, "Impact of climate change on extreme rainfall events and flood risk in India", *J. Earth Syst. Sci.,* vol. 120, no. 3, pp. 359-373, 2011.
[http://dx.doi.org/10.1007/s12040-011-0082-5]

[27]   N. Saith, and J. Slingo, "The role of the Madden–Julian Oscillation in the El Niño and Indian drought of 2002", *Int. J. Climatol.,* vol. 26, no. 10, pp. 1361-1378, 2006.
[http://dx.doi.org/10.1002/joc.1317]

[28]   T. Sato, F. Kimura, and A. Kitoh, "Projection of global warming onto regional precipitation over Mongolia using a regional climate model", *J. Hydrol. (Amst.),* vol. 333, no. 1, pp. 144-154, 2007.
[http://dx.doi.org/10.1016/j.jhydrol.2006.07.023]

[29]   R. Selvaraju, "Impact of El Niño-southern oscillation on Indian foodgrain production", *Int. J. Climatol.,* vol. 23, no. 2, pp. 187-206, 2003.
[http://dx.doi.org/10.1002/joc.869]

[30]   C.W. Thornthwaite, and J.R. Mather, "The water balance", In: *Climatology.Drexel Institute of Technology, Laboratory of Climatology.* vol. 8. Centerton NJ, 1955, no. 1, p. 104.

[31]   S. Venkataraman, and A. Krishnan, "Crops and Weather, Publica-tions and Information Division", In: *Indian Council of Agricultural Research.* New Delhi, 1992, p. 586.

[32]   Y.C. Hountondji, F. de Longueville, and P. Ozer, "Trends in extreme rainfall events in Benin (West Africa), 1961-2000", *Proceedings of the 1st international conference on energy. Environ. Clim. Change 7. August 26-27, 2011,* Ho Chi Minh City Vietnam., 2011.http://hdl.handle.net/2268/96112

[33]   D.K. Panda, A. Mishra, A. Kumar, K.G. Mandal, A.K. Thakur, and R.C. Srivastava, "Spatiotemporal patterns in the mean and extreme temperature indices of India, 1971-2005", *Int. J. Climatol.,* vol. 34, no. 13, pp. 3585-3603, 2014.
[http://dx.doi.org/10.1002/joc.3931]

[34]   S.K. Jain, K. Vijay, and M. Saharia, "Analysis of rainfall and temperature trends in northeast India"", *Int. J. Climatol.,* 2012.
[http://dx.doi.org/10.1002/joc.3483]

[35]   A. Kitoh, M. Hosaka, Y. Adachi, and K. Kamiguchi, "Future projections of precipitation characteristics in East Asia simulated by the MRI CGCM2", *Adv. Atmos. Sci.,* vol. 22, no. 4, pp. 467-478, 2005.
[http://dx.doi.org/10.1007/BF02918481]

[36]   R.K. Singh, S.J. Kumar, A. Singh, R. Raju, and D.K. Sharma, "Adaptation in rice-wheat based sodicagroecosystems: A case study on climate resilient farmerts' practices", *Indian J. Tradit. Knowl.,* vol. 12, no. 1, pp. 377-389, 2014.

[37]   L.N. Joppa, B. O'Connor, P. Visconti, C. Smith, J. Geldmann, M. Hoffmann, J.E.M. Watson, S.H.M. Butchart, M. Virah-Sawmy, B.S. Halpern, S.E. Ahmed, A. Balmford, W.J. Sutherland, M. Harfoot, C. Hilton-Taylor, W. Foden, E.D. Minin, S. Pagad, P. Genovesi, J. Hutton, and N.D. Burgess, "Filling in biodiversity threat gaps", *Science,* vol. 352, no. 6284, pp. 416-418, 2016.
[http://dx.doi.org/10.1126/science.aaf3565] [PMID: 27102469]

[38]   P. Dhanya, A. Ramachandran, and P Prasanthkumarbal, *Adaptation in rice-wheat based sodicagroecosystems: A case study on climate resilient farmerts' practices".* Indian J Trad Knowledge., vol. 12, no. 1, pp. 377-389, 2013.

[39]   J. Sheffield, and E.F. Wood, "Projected changes in drought occurrence under future global warming from multi-model, multi-scenario, IPCC AR4 simulations", *Clim. Dyn.*, vol. 31, no. 1, pp. 79-105, 2008.
[http://dx.doi.org/10.1007/s00382-007-0340-z]

[40]   M.V.K. Sivakumar, *Interactions between climate and desertification.* vol. 142. Agric Forest Meterol, 2007, pp. 143-155.

[41]   S.V.R.K. Prabhakar, *Climate change adaptation implications for drought risk mitigation: a perspective for India.,* 2008.

[42]   R. Shaw, S.V.R.K. Prabhakar, and A. Fujieda, *Community level climate change adaptation and policy issues: a case study from Gujarat, India.* Graduate School of Global Environmental Studies: KyotoUniversity, Japan, 2005, p. 59.

[43]   B. Smit, and M. Skinner, *Adaptation options in agriculture to climate change: a typology.* vol. Vol. 7. Mitigation and Adaptation Strategies for Global Change, 2002, pp. 85-114.

[44]   W. Neil Adger, N.W. Arnell, and E.L. Tompkins, "Successful adaptation to climate change across scales", *Glob. Environ. Change,* vol. 15, no. 2, pp. 77-86, 2005.
[http://dx.doi.org/10.1016/j.gloenvcha.2004.12.005]

[45]   A.J. Challinor, T.R. Wheeler, P.Q. Craufurd, C.A.T. Ferro, and D.B. Stephenson, "Adaptation of crops to climate change through genotypic responses to mean and extreme temperatures", *Agric. Ecosyst. Environ.,* vol. 119, no. 1-2, pp. 190-204, 2007.
[http://dx.doi.org/10.1016/j.agee.2006.07.009]

SUBJECT INDEX

G. Venkatesan, S. Lakshmana Prabu and M. Rengasamy (Eds.)

O

Oil(s) 7, 15, 41, 42
  palm 15
  refinery 7
  seed 42
Operation, wastewater treatment 166
Organic 90, 91, 92
  contaminant degradation 91, 92
  hydrocarbons 90
Oxidation 77, 79, 85, 167
  chemical 167
  thermal 167
Oxygen 77, 85, 89, 108
  dissolved 77

P

Packaging 27, 160
  biodegradable 160
Pesticides 53, 90, 93
Photo-Fenton process 85
Phreatophyte trees 91, 94, 95
Physical-chemical 74, 80
  effluent treatment techniques 80
  methods 74
Physico-chemical methods 90
Phytodegradation 91, 92
Plant(s) 16, 17, 18, 24, 28, 30, 42, 74, 75, 77,
    90, 91, 92, 93, 94, 95, 96, 155, 159, 170,
    196
  automobile assembly 155
  community wastewater 159
  growing 24
  metabolism 95
  microbes 91
  renewable organic 42
  sewage treatment 196
  wastewater treatment 170
  water treatment 24
Pollutant(s) 25, 74, 78, 79, 81, 85, 87, 88, 89,
    90, 91, 93, 94, 97, 150, 151, 167, 168,
    170, 179
  absorbed 94

biodegradable 179
concentration 89
corrosive 167
detoxification 90
industrial 74, 97
organic 85, 91, 93
suspended solid 81
Pollution 1, 2, 13, 20, 24, 58, 77, 97, 113, 114,
    118, 148, 149, 151, 155
  generation 24, 77
  noise 58
  preventing 2
  secondary 97
Pressure 1, 9, 53, 137, 173, 175, 187
  environmental 53
  growing 9
  osmotic 187
Principles 1, 2, 3, 11, 12, 16, 18, 19, 20, 21,
    23, 24, 25, 46
  adopted 46
  natural 20
Process 3, 8, 9, 15, 17, 20, 21, 22, 24, 25, 28,
    60, 80, 84, 85, 86, 90, 91, 92, 94, 95, 96,
    106, 108, 109, 113, 139, 174, 199, 214,
    215
  aridification 215
  aridity 199, 214
  automated 60
  decision-making 113
  endothermic 139
  exothermic 174
  filtration 80
  hydrological 215
  ozonation 86
  phytodegradation 91
  phytoextraction 92
  phytoremediation 90, 95, 96
Production 1, 16, 200
  agricultural 200
  industrial 16
  of edible mushroom 16
  sustainable 1
Products 1, 2, 3, 6, 8, 9, 11, 12, 13, 15, 19, 20,
    21, 22, 24, 31
  human industry 21

www.ingramcontent.com/pod-product-compliance
Lightning Source LLC
Chambersburg PA
CBHW050830220326
41598CB00006B/348